The Chemistry of Plants and Insects

Plants, Bugs, and Molecules

The Chemistry of Plants and Insects

Plants, Bugs, and Molecules

Margareta Séquin

Department of Chemistry and Biochemistry,
San Francisco State University, USA
Email: msequin@sfsu.edu

THE QUEEN'S AWARDS
FOR ENTERPRISE:
INTERNATIONAL TRADE
2013

All photos are by Margareta Séquin, unless credited otherwise in the figure captions.

Print ISBN: 978-1-78262-448-6

A catalogue record for this book is available from the British Library

The Royal Society of Chemistry is a charity, registered in England and Wales, Number 207890, and a company incorporated in England by Royal Charter (Registered No. RC000524), registered office: Burlington House, Piccadilly, London W1J 0BA, UK, Telephone: +44 (0) 207 4378 6556.

Visit our website at www.rsc.org/books

Printed in the United Kingdom by CPI Group (UK) Ltd, Croydon, CR0 4YY, UK

Preface

When we observe plants in the outdoors, we are likely to notice the visiting insects. Plants and insects communicate in numerous ways, and chemistry plays a key role in these communications. Natural organic compounds determine whether a plant is consumed by insects or avoided by them, and which insects may pollinate its flowers. I was intrigued by these connections during a field seminar where the task was to search for insects. While I found few insect samples initially, the entomologists in the group collected plenty of exciting beetles, caterpillars, and butterflies. My colleagues clearly knew to look for distinct host plants whose plant chemistry attracted specific insects.

This book is aimed at lay readers who have a basic understanding of chemistry. It addresses non-chemists who work at, or enjoy visiting, botanical gardens and science centers, and non-major undergraduate college students. The book can be a supplementary text for students in plant sciences, ecology, entomology, and horticultural programs. By connecting chemistry with plants and insects, the author intends to capture the interest of readers who would like a deeper understanding of the natural world.

Many people have supported me in the writing of this book. I am most grateful to the readers and commentators who generously contributed their suggestions and their knowledge during the book's preparation. I am especially grateful to Jim Keeffe, Eileen Nottoli, Urs Séquin, Verena Rau, and Eveline Larrucea for their encouragement and for their detailed critical reading of the book chapters. Many thanks to John Hafernik for a wonderful field trip and for checking the insect entries. Any errors that stubbornly resisted detection are

The Chemistry of Plants and Insects: Plants, Bugs, and Molecules
By Margareta Séquin
© Margareta Séquin 2017
Published by the Royal Society of Chemistry, www.rsc.org

entirely the author's responsibility. I am grateful to Sandy Jordan for providing insect samples, to Sushila Kanodia for samples of neem, and to Christa Kraus for helpful information on honey and beeswax. My special thanks go to the staff at the Regional Parks Botanic Garden in Berkeley, CA, and to the staff of the University of California Botanical Garden who alerted me to special plants and their insects.

I am greatly indebted to the team at the Royal Society of Chemistry for their continued support and excellent service.

My special thanks go to my husband Carlo for his support, encouragement, and patience.

And many thanks to Elise, Sienna, and Laila for supplying bug materials.

Margareta Séquin

Caterpillar topiary at San Francisco State University.

Contents

The Chemistry of Plants and Insects: Plants, Bugs, and Molecules
By Margareta Séquin
© Margareta Séquin 2017
Published by the Royal Society of Chemistry, www.rsc.org

3   Plants That Eat Insects   48

4   Plants' Defense Against Insects   54

Part 2: The Insect Perspective

5   Insects and Their Chemistry   83

Part 3: Plants and Insects: The Human Perspective

Introduction

1 The Chemistry of Plants and Insects

This book is an introduction to the natural chemical compounds that are part of the many different interactions between plants and insects. Fragrant scents from flowers attract pollinators. Bright or drab pigments in flower petals tempt different kinds of pollinating insects to visit. Flowers with stamens full of pollen provide food to honey bees, and sweet sugary nectar offerings in flowers invite butterflies, all as an enticement to encourage pollination and, with this, to promote reproduction of the plants (Figure 1.1). Organic, *i.e.* carbon-based, chemical compounds compose the floral scents, pigments, and nectar components that perform key roles in the communications between plants and insects.[1,2]

Most insects, on the other hand, use and need plants as vital sources of food, in particular to obtain the basic nutrients of carbohydrates, amino acids, and fatty acids. Insects and other animals (including humans) are *heterotrophs* and as such cannot synthesize all the *primary metabolites* in their own systems but must obtain them from plant-related sources. (Refer to the Glossary at the end of this book for definitions and brief explanations if needed.) Huge numbers of beetles, caterpillars, grasshoppers, and aphids find their basic nutrients in tender plant parts (Figure 1.2(a)). Some insects obtain them from plant roots, tree barks, or fruits while bees, butterflies, and many beetles find nourishing proteins and sugars in pollen or nectar from flowers. There are, of course, insects that feed on other animals, like mosquitoes sucking blood, or on animal products, like dung beetles using animal feces as food (Figure 1.2(b)). But these insects

The Chemistry of Plants and Insects: Plants, Bugs, and Molecules
By Margareta Séquin
© Margareta Séquin 2017
Published by the Royal Society of Chemistry, www.rsc.org

(a)   **(b)**

Figure 1.1   Plants attracting insect pollinators. (a) Bright floral pigments and nectar of baby blue eyes flowers (*Nemophila menziesii*) attract a honey bee (*Apis mellifera*). (b) A swallowtail butterfly (*Papilio machaon*) is lured by the floral color, nectar, and pollen of a thistle (*Carduus* sp.).

(a)   **(b)**

Figure 1.2   Insects feeding on plants or animal products. (a) Cockchafer or May beetle (*Melolontha* sp.), a voracious European herbivore. (b) African dung beetle (*Circellium* sp.) rolling dung ball. (Photo by Greg H. Rau.)

get their basic nutrients from animals that had earlier ingested plant foods and with them the required primary metabolites.[3]

Plants, as *autotrophs*, are able to undergo photosynthesis. They can convert carbon dioxide from air and water into oxygen gas and simple sugars like glucose, in the presence of chlorophyll, light, enzymes, and mineral ions. From the simple sugars, plants can synthesize larger carbohydrates, amino acids, and fatty acids, as well as vitamins. All are basic nutrients that need to be ingested by insects and other animals. In further biosynthetic steps, plants can produce attractive scents and colorful pigments that call out to pollinators.

Plants not only produce a wealth of organic compounds that serve to attract pollinators, but also synthesize highly diverse defensive compounds. Strong smells in leaves, bitter substances in leaves or roots, and toxic compounds in fruits, leaves, or roots, fend off insects that would otherwise eat the plants and destroy them. Aside from many physical defenses like thorns, prickles, and tough skins, these defensive substances protect plants from being devoured by insects keen on feeding on leaves, roots, or heartwood. Tender plant shoots with low concentrations of natural defensive chemicals are most susceptible to insect attacks. So are plants that are stressed by environmental factors and so are less capable of synthesizing compounds that are distasteful or toxic to insects.

Unlike most plants, insects are short-lived, from a few days to a couple of months.[4] Beetles, caterpillars, aphids, and other insects can produce generation after generation in large numbers and are able to rapidly evolve to adapt to the chemistry of the plants they visit. Many insects have developed a tolerance and even a preference for previously unpalatable plant sources. An example is the Colorado potato beetle (*Leptinotarsa decemlineata*), infamous for its damaging infestations on potato plants (Figure 1.3(a)).[5] The beetles most likely had their original ranges in Mexico and Southwestern North America, feeding on some native plants there. With the development

(a)   (b)

Figure 1.3 Chemical plant defenses and insects. (a) Colorado potato beetle (*Leptinotarsa decemlineata*) feeding on a potato plant (*Solanum tuberosum*). (Photo: Wikimedia Commons. https://commons.wikimedia.org/wiki/File:Colorado_potato_beetle.jpg, (accessed September 2016).) (b) Red milkweed beetle (*Tetraopes femoratus*) feeding on milkweed plant (*Asclepias* sp.), ingesting toxins from the plant latex.

of cultivated potatoes (*Solanum tuberosum*) in North America, the beetles adapted to feeding on the cultivated crops as their host plants. Through the export of potato plants, the voracious insects were soon further introduced to temperate climates all over the world. Rapid cycles of generations enabled the potato beetles to adapt and infest the crops in spite of toxic chemical defenses in the plants.[6]

Numerous species of insects have not only learned to tolerate plant defensive compounds without harm, but use the toxins for their own protection from predators. For example, the caterpillars of monarch butterflies (*Danaus plexippus*), as well as specialized aphids and beetles like the red milkweed beetle shown in Figure 1.3(b), have adapted to feed on milkweed plants (*Asclepias* spp.) and can ingest defensive plant compounds from the plants' milky latex without harm. The insects then use the plant toxins for their own defense, becoming ill-tasting and unpalatable to their predators.[7] Insects have even evolved mechanisms to slightly alter the chemical structures of defensive plant compounds, transforming them to pheromones that attract more insects. The *coevolution* of insects and plants continues, and crucial organic compounds, produced by the respective plants and ingested and occasionally tolerated by feeding insects, play key roles.[8]

The examples of insect–plant interactions chosen in this book and the chemical compounds that participate in them are necessarily a very limited selection from the huge number of such relationships. The case studies here are chosen from around the world and selected for their fairly well-known roles, be they desirable or undesirable to mankind, or because of the particular interest of their organic compounds. The great diversity of chemical compounds that are part of these interactions will serve as a presentation of the major families of organic compounds occurring in living things. As the book progresses, the selected sequence of topics and the related examples of chemical compounds will gradually proceed from the relatively simple organic structures of plant volatiles, like geraniol **1.1** (Figure 1.4(a)) found in scents of flowers, to increasingly more complex molecules, like the toxic alkaloid solanidine **1.2** (Figure 1.4(b) found in potatoes.

The book begins with the plant perspective, with an introduction to the structures of organic compounds that plants synthesize in order to communicate with insects. Many components of fragrances in flowers are relatively small organic molecules (on a molecular scale), with few carbon atoms assembling them. Their structures will serve as

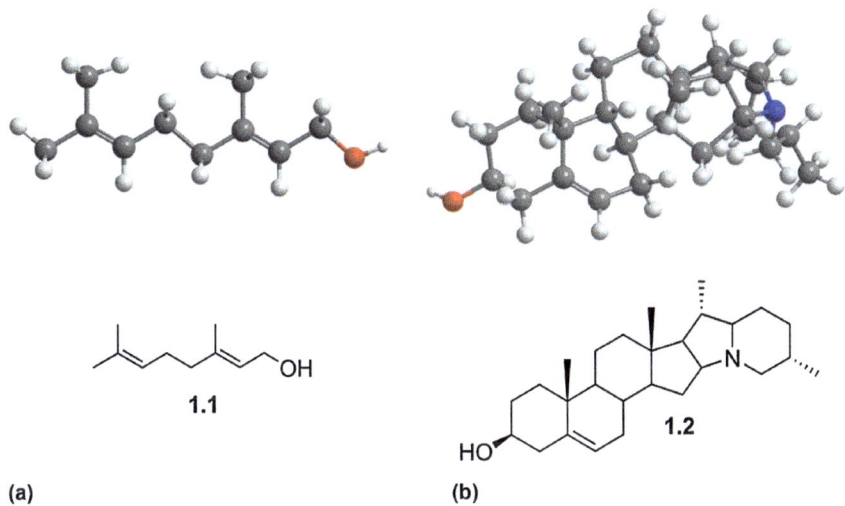

Figure 1.4   The chemical structures of (a) geraniol **1.1**, a fragrant, oily compound found in rose oil, and (b) solanidine **1.2**, a toxic compound found in potato sprouts, are both shown as molecular structures and as simplified line structures.

an introduction to typical organic molecules. The composition of flower nectars and their sugars will introduce additional chemistry concepts. We will then progress to the more complex molecules of colorful pigments, with continued reflections on how the plant compounds influence insect responses. Pollen in flowers is not only transferred by insects onto stigmas of other flowers, inducing fertilization of the plants; it is also offered to attract insects as a nutritious source of starch and proteins, examples of natural polymers. Some plants, namely the insectivorous plants, lure insects to intricately-shaped modified leaves, trap the insects, and then use special enzymes to digest them and use the digestion products as nutritional supplements. The subsequent chapter describes the great diversity of plant compounds that act as insect repellents. It further illustrates the wealth of organic plant compounds that are important in interactions with insects.

In the chapters of Part 2, previously introduced families of organic compounds will reappear, but as part of the insect perspective. The focus will be first on the special chemistry of insects themselves, and then on the nutrients that insects obtain from eating plants. Some insects, like many types of aphids (*Aphis* spp.), are generalists that feed on lots of different types of plants (Figure 1.5(a)). Others, like the

(a)   (b)

Figure 1.5 Generalist insects and specialists. (a) Many types of aphids are generalists, here infesting the vegetable Swiss chard (*Beta vulgaris* ssp. *vulgaris*). (b) An elderberry longhorn beetle (*Desmocerus californicus*) feeding on elderberry leaves (*Sambucus* sp.), its host plant. (Photo by Eveline Larrucea.)

elderberry beetle (*Desmocerus californicus*, Figure 1.5(b)), are selective about the plants they feed on. These choices are based on the chemistry of the respective plants. Numerous insects use plant defenses for their own defense. Insects that can ingest plant toxins without harm often advertise their acquired toxicity with bright warning colors and distinct patterns, as illustrated in Figures 1.3(a) and (b), and 1.5(b).

Comparing plants and insects with regards to the composition of their structural materials, their pigments, or their defensive substances is fascinating. The respective chapter sections will include comparisons of the organic compounds that compose them.

Interactions between plants and insects and their ecology are of key interest to mankind, and the chapters in Part 3 concentrate on the human dependence on insect–plant communications. People would go hungry without the work of pollinating insects in orchards and on other food crops. Some insect products, like honey, beeswax, and silk, are of great interest to humans. They are products of specific insects feeding on plant sources. On the other hand, insects can appear as destructive pests on crop plants, threatening the supply of food for humans. Therefore, studies on insect attractants and repellents, as well as genetic studies on how to breed plants that are less susceptible to insect infestations, are active topics of research. Various approaches on how to manage insect pests, including thoughts on environmentally sensible methods, will be discussed.

Plant–insect interactions comprise a field of ongoing and vigorous research. Previously unknown insects and plants keep being discovered, especially in environments like tropical rain forests, and with them new mechanisms of communications between plants and insects. Often the amounts of essential chemical compounds that are involved in specific interactions are extremely small and therefore difficult to detect or to define. Thanks to increasingly sensitive methods of instrumentation, previously unknown mechanisms and their participating chemical compounds have been elucidated, and great recent progress has been made in explaining them.[9,10] Examples of investigations and their findings will be shown.

The subtitle of this book is, somewhat tongue-in-cheek, "Plants, Bugs, and Molecules", the word 'bugs' used here for insects in general, just as in everyday language. In *entomology*, *i.e.* the scientific studies of insects, true bugs represent a subcategory, called an order, of the class of insects. True bugs are scientifically known as the order of the Hemiptera. We shall use the scientific definition in this book. Many 'bugs' from everyday language are actually beetles (order Coleoptera), like ladybugs which are more correctly called lady beetles or ladybird beetles. True bugs have very different mouthparts from beetles. As another example of a striking difference between bugs and beetles, the larval stages of beetles, like the lady beetles shown in Figure 1.6(a) and (b), look significantly different from the adult insects. Beetles go through a complete metamorphosis during their development, whereas true bugs, like the milkweed bugs shown in Figure 1.6(c), look quite similar in their immature stages and as adults. More details on insect and plant biology will be shown in the respective chapters if they are relevant to insect–plant interactions and their chemistry.

There is a wealth of common names for individual plants and insects, and many local variations exist aside from language differences. On the other hand, the systematic names for individual plants or insects are unique and internationally used. Therefore, we'll include systematic names after common names of plants or insects throughout the book. Thus, a ladybug would be referred to as ladybug or lady beetle (*Coccinella californica*) (Figure 1.6(a)).

Refer also to the Glossary at the end of this book for definitions and brief explanations, as well as for a short guide to reading structures of organic compounds.

(a)   (b)   (c)

Figure 1.6   Beetles and bugs. (a) California ladybug or ladybird beetle (*Coccinella californica*) as an adult. (b) A ladybird beetle larva on a raspberry. (c) Adult milkweed bug (*Oncopeltus fasciatus*) with young bugs. (Photo by Greg Hume. https://commons. wikimedia.org/wiki/File:Oncopeltusfasciatus.jpg#/media/ File:Oncopeltusfasciatus.jpg, (accessed September 2016).)

Bibliography and Further Reading

The end-chapter references include books as well as review articles and journal articles that relate to the topics of the respective chapter. While most referenced books and articles are in-depth texts, others are for popular reading.

Many excellent texts can provide detailed background on organic chemistry, biochemistry, plant biology, and entomology if needed.

Some examples are:

J. E. McMurry, *Fundamentals of Organic Chemistry*, Brooks/Cole, Belmont, CA, 8th edn, 2012.

D. R. Klein, *Organic Chemistry*, J. Wiley and Sons, 2nd edn, 2013.

R. F. Evert and S. E. Eichhorn, *Raven Biology of Plants*, W. H. Freeman, New York, NY, 8th edn, 2012.

D. L. Nelson and M. M. Cox, *Lehninger Principles of Biochemistry*, W. H. Freeman, New York, NY, 6th edn, 2012.

H.-W. Heldt and B. Piechulla, *Plant Biochemistry*, Academic Press, London, 4th edn, 2011.

P. J. Gullan and P. S. Cranston, *The Insects: An Outline of Entomology*, Wiley-Blackwell, Chichester, West Sussex, UK, 4th edn, 2010.

R. J. Elzinga, *Fundamentals of Entomology*, Pearson, Upper Saddle River, NJ, 6th edn, 2004.

J. B. Harborne, *Introduction to Ecological Biochemistry*, Academic Press, London, 4th edn, 1993.

For reference on organic structures:

The Merck Index: An Encyclopedia of Chemicals, Drugs, and Biologicals,
ed. M. J. O'Neil, The Royal Society of Chemistry, Cambridge, UK,
15th edn, 2013.

References

1. M. Séquin, *The Chemistry of Plants: Perfumes, Pigments, and Poisons*, Royal Society of Chemistry, Cambridge, UK, 2012.
2. P. Willmer, *Pollination and Floral Ecology*, Princeton University Press, Princeton and Oxford, 2011.
3. L. M. Schoonhoven, J. J. A. van Loop and M. Dicke, *Insect-Plant Biology*, Oxford University Press, Oxford, 2005.
4. R. J. Elzinga, *Fundamentals of Entomology*, Pearson, Upper Saddle River, NJ, 6th edn, 2004.
5. J. D. Hare, Ecology and management of the Colorado potato beetle, *Annu. Rev. Entomol.*, 1990, **35**, 81.
6. J. R. Hanson, *Chemistry in the Garden*, Royal Society of Chemistry, Cambridge, UK, 2007.
7. T. Eisner, *For Love of Insects*, Harvard University Press, Cambridge, MA, 2003.
8. J. B. Harborne, *Introduction to Ecological Biochemistry*, Academic Press, London, 4th edn, 1993.
9. R. Raguso, W. Boland, T. Hartmann, J. A. Pickett and D. Strack, Editorial: Plant-insect interactions, *Phytochemistry*, 2011, **72**, 1495.
10. H. M. Schaefer and G. D. Ruxton, *Plant-Animal Communication*, Oxford University Press, Oxford, 2011.

Part 1: The Plant Perspective

2 Plants Attracting Insects

2.1 Introduction

Survival and successful reproduction are essential for the continuation of life for all living things. As for plants, reproduction greatly depends on insect pollination. In about two thirds of all the species of flowering plants, known as the *angiosperms*, insects pollinate the flowers by moving pollen from the male anthers of one flower to the female stigma of another flower of the same species. (Figure 2.1 shows a schematic drawing of the basic parts of a flower.) Pollination induces fertilization of the plant, followed by seed production, and thus promotes reproduction of the species. Organic compounds produced by the plants have major roles in advertising to potential pollinators and attracting them. Many plants feature colors in their flowers that are visible and appealing to insect pollinators (Figure 2.2(a) and (b)). In addition, flowers emit attractive scents when ready for pollination. Many flowers also offer sweet nectars and ample pollen as food, to further entice insects to visit.

Fossil finds of plants and insects have shown that towards the end of the early Cretaceous era, between 130 and 90 million years ago, a rapid evolution of plants and insects occurred. Diversification of plants, with fast-increasing numbers of plant species, took off when insects started to evolve during that time period. It is generally assumed that it was the diversification of nectar- and pollen-collecting insects that greatly drove the evolution of plants.[1]

Compared to pollination by insects, pollination by wind is the more ancient form in evolutionary terms. Examples of wind-pollinated

The Chemistry of Plants and Insects: Plants, Bugs, and Molecules
By Margareta Séquin
© Margareta Séquin 2017
Published by the Royal Society of Chemistry, www.rsc.org

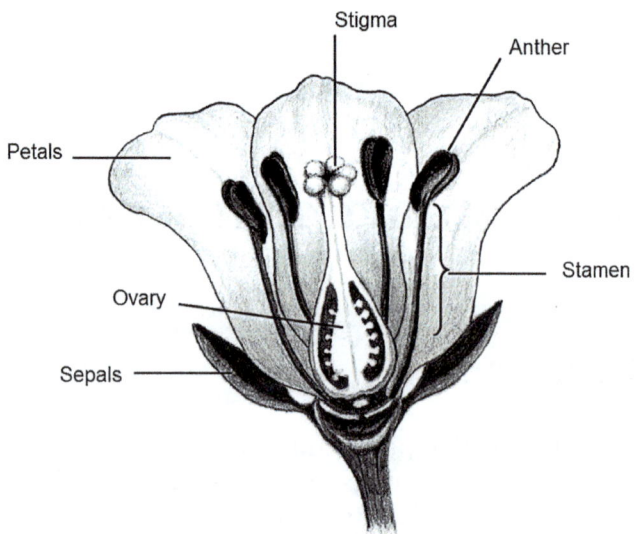

Figure 2.1   Schematic drawing of the flower parts of a flowering plant. (Drawing by Eveline Larrucea.)

(a)   (b)   (c)

Figure 2.2   Insect-pollination *versus* wind-pollination. (a) Colorful flowers with plenty of pollen, like the blossoms of a beavertail cactus (*Opuntia basilaris*), attract pollinating insects. (b) Brightly-colored flower petals, with nectar guides and floral nectar, attract beetle pollinators in a yellow mariposa tulip (*Calochortus luteus*). (c) Wind-pollinated plants, like live oaks (*Quercus* sp.), have inflorescences with inconspicuous colors and produce large amounts of pollen.

plants are pines (*Pinus* spp.), oaks (*Quercus* spp., Figure 2.2(c)), and all the grasses (Poaceae). Wind-pollinated plants tend to have drab floral parts and no nectar. This type of pollination obviously requires wind and also masses of pollen which is randomly dispersed into the surrounding space. Insect pollination, on the other hand, requires smaller quantities of pollen for successful fertilization. Insects seek out plants that have the proper attractive features in the form of enticing colors and smells, aside from inviting floral shapes, and tend to be quite faithful to the particular plant species they visit. Efficient mixing of plant genes takes place as insects transport male genetic information contained in pollen to the female floral parts of another flower of the same species. Self-pollination in insect-pollinated flowers is uncommon. In contrast to dispersal of pollen by wind, pollination by insects can successfully occur in scattered and isolated plant populations, as can be found in tropical forests. Specific floral scents and attractive pigments direct insects to isolated plants of the same species at considerable distances, and pollinating insects can detect and visit the matching plants.

2.2   Plant Volatiles That Attract Pollinating Insects

Plants produce large numbers of *volatile* compounds, *i.e.* compounds that easily evaporate into the air. More than 1700 different volatile organic compounds, sometimes abbreviated as VOCs, have been identified in flowers alone.[2] We notice these volatiles if they have distinct odors to us, pleasant or unpleasant, especially on a warm day. (Note that the characterizations of the nature of the smells are necessarily from a human point-of-view.) Scents emitted by flowers alert potential pollinators. Because plants are mostly rooted in one place, the volatiles are a means of long-distance communication with other organisms, like insects. Compared to visual cues from flowers such as colors, odorous volatiles are randomly sent into the surrounding space and thus have low directionality. Concentrations of the scents emitted by flowers are affected by wind conditions and can be disturbed by smoke, as from forest fires, and other pollutants.[3] The scents, together with appropriate floral colors and shapes of the flowers, are inviting to select pollinators.

   Some flowers attract a broad range of pollinators, whereas others only appeal to a few types of insects. Sweet scents, like the strong fragrance of the damask rose (*Rosa damascena*) (Figure 2.3(a)), are generally attractive to butterflies, bees, and bumble bees. Moths are lured by heavy, aromatic smells of flowers that tend to open later in the day, like the blossoms of white jasmine (*Jasminum* sp.) (Figure 2.3(b)). Maroon-colored or brown inflorescences often produce smells of decaying fish or of carrion, and are attractive to flies or small gnats. Floral smells are emitted during specific times of a

(a)   (b)   (c)

Figure 2.3   Fragrant and foul-smelling flowers. (a) Damask rose (*Rosa damascena*) is a highly fragrant, sweet-smelling rose. (b) Heavy-scented jasmine flowers (*Jasminum* sp.) release their fragrance late in the day, attracting moths. (c) A large inflorescence of the corpse plant (*Amorphophallus titanum*), about two meters tall, is ready to open and to release foul-smelling volatiles that attract flies.

plant's growth phase, such as during the opening of flowers. Some plants even enhance the evaporation of attractive compounds, heating up during the flowering phase by increasing metabolic rates. This phenomenon is especially common in plants of the Arum family. A famous example is the Sumatran corpse plant (*Amorphophallus titanum*, Figure 2.3(c)) whose giant inflorescence can heat up by more than 10 °C, sending out volatiles with a rotten-meat smell that attracts fly pollinators.[4]

We shall encounter volatile compounds again in a later chapter, but there the volatiles are emitted not from flowers but from plant parts like leaves and needles. Under these circumstances, the volatiles can act as deterrents towards insects and other animals, keeping the browsers from feeding on the plants and thus supporting the plants' survival.

2.3   Compositions of Plant Scents

Plant scents are mixtures of sometimes more than a hundred different volatile organic compounds. These compounds are generally oily liquids at ambient temperatures. This property gave them the general name *essential oils*, 'essential' here meaning the characteristic or essence of a scent. The composition of a plant scent is influenced by the type of plant, its age, its growth phase, and the time of day, as well as the local climate and the soil conditions. These factors also determine the concentrations, *i.e.* the percentages, of compounds found in a particular fragrance.

If we want to determine the components of a floral scent, the most commonly used method is gas chromatography (GC) combined with mass spectrometry (MS). Gas chromatography is a highly sensitive analytical method that is used to separate mixtures of volatile compounds, requiring only fractions of a microliter (*i.e.* 10^{-6} liters) for analysis. It is therefore a most valuable method to separate and examine mixtures of volatiles for which only very small quantities are available. This certainly applies to essential oils in floral fragrances of which often only trace amounts are present.[5]

In order to study the components of a fragrance by GC, we first need to collect a sample of the scent mixture. A frequently-used, gentle method is head-space analysis. Figure 2.4 shows a head-space analysis set-up for collecting volatiles emitted by the fragrant flowers of a ylang ylang shrub (*Cananga odorata*). A glass bell jar is placed over the fragrant flowers and connected with a probe containing a material that can absorb volatile oils. The floral volatiles are pulled with a pump into the probe. This method is non-destructive to the plant parts. The collected oils are eluted, *i.e.* rinsed, from the probe with appropriate solvents and can then be further separated and analyzed by gas

Figure 2.4 Head space analysis set-up to collect volatiles emitted by the fragrant flowers of an ylang shrub (*Cananga odorata*), Longwood Gardens, Pennsylvania.
(Photo Courtesy Longwood Gardens.)

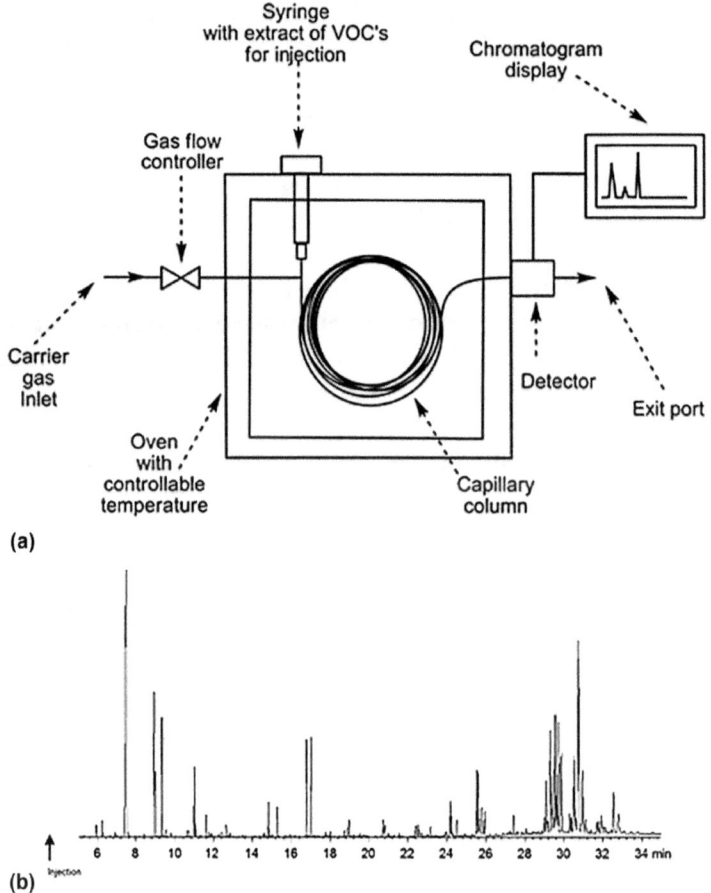

Figure 2.5   Analysis of a mixture of volatile compounds, as in a floral scent, by gas chromatography. (a) Schematic drawing of a gas chromatograph. (b) A sample gas chromatogram of a mixture of volatile oils.
(Gas chromatogram by Phenomenex, Inc.)

chromatography (Figure 2.5). If combined with mass spectrometry (MS), the identity of the single compounds in a mixture can be determined by GC-MS. Figure 2.5(a) shows a schematic picture of a gas chromatograph. Figure 2.5(b) displays an example of a gas chromatogram of a sample of a mixture of volatile oils, as in a floral scent, with each line in the chromatogram representing a different compound.[6]

In a related method, used to evaluate the odor contributions of the compounds in a volatile mixture, a 'sniffing port' is attached to the exit port of the gas chromatograph. Each compound can then be smelled as it is being separated and can be further evaluated for its individual odor quality. Humans have no odor perception of some compounds in the volatile mixtures of plant scents, but these may well be recognized by animals like insects. (After all, plants did not evolve floral scents to attract humans.) The sensitivity towards different volatiles can be very different among insects, and a higher percentage of a volatile compound in the mixture does not necessarily mean that it is more easily perceived. On the other hand, volatiles present in trace amounts can be crucial scent components.

Table 2.1 lists a few of the more than two hundred volatile compounds found in rose oil, extracted from the fragrant damask rose (Figure 2.3(a)), with their percentage in the extract and a description of their aroma. The listed compounds will be part of our subsequent discussions where their molecular structures will introduce the chemistry of plant volatiles and the families of organic compounds they belong to.

Molecules that compose plant volatiles have common chemical characteristics in spite of their structural diversity. They are organic molecules of relatively small sizes (on a molecular scale), each containing up to about fifteen carbon atoms. The molecules are mostly *hydrocarbon* structures, *i.e.* hydrogen and carbon atoms compose their molecular structures. They are *lipophilic* (fat-loving) and *hydrophobic* (water-repelling), meaning they are insoluble or only slightly

Table 2.1 %-Composition of some volatile compounds in an extract of the essential oils of *Rosa damascena*, as detected in a sample collected by head-space and analyzed by GC-MS.[a]

Compound	%-Concentration	Aroma impressions
α-Pinene	2.48	Woody, pine-like
Myrcene	1.01	Sweet-balsamic
Limonene	0.22	Fresh, fruity, citrus
1,8-Cineole	0.03	Fresh, eucalyptus-like
Acetic acid	0.16	Powerful sour
Linalool	2.12	Floral
Citronellol	37.05	Floral, intense rose-like
Nerol	9.91	Floral rose-like
Geraniol	18.62	Rose-like, sweet floral
2-Phenylethanol	2.94	Floral, rose-like, honey notes
Farnesol	0.18	Floral-oily

[a]L. Jirovetz, G. Buchbauer, A. Stoyanova, A. Balinova, Z. Guangjiun and M. Xihan, Solid phase microextraction/gas chromatographic and olfactory analysis of the scent and fixative properties of the essential oil of *Rosa damascena* from China, *Flavour Fragrance J.*, 2005, **20**(1), 7.

soluble in water. There is little attraction, or *intermolecular bonding*, between the molecules of volatile plant scents. As a consequence, compounds composed of these molecules evaporate rapidly, especially at elevated temperatures. The volatiles enter the gas phase and travel through the air, reaching the sensors of smell in insects or other animals (including humans). *Functional groups* may be attached to the hydrocarbon structures of volatiles, like alcohol groups (OH), aldehyde (CHO), or ester groups (COOR). They contribute to characteristic aroma notes of the respective compounds.

Plants synthesize the volatile compounds from basic primary metabolites, like simple carbohydrates and amino acids, following defined biochemical reaction pathways. The components of plant scents, as well as of plant pigments and of the plant defensive compounds that we shall encounter later, are collectively known as plant *secondary metabolites*; they are organic compounds that are not directly involved in the normal growth, development, or reproduction, and are not found in every plant species, but that nevertheless have important functions in plants.

Although there are large numbers of different plant volatiles, there is a lot of recurrence of the same compounds in the various plant scents. But they are found in very different percentages in the scent mixtures. This allows for a huge variety and for endless possibilities of floral smells. Insects like honey bees detect these differences and seek out specific flowers. The following descriptions of structures of volatiles will address compounds that are common in floral scents.

2.4   Ancient Plants, Hydrocarbons, and Beetle Pollination

For our introduction to organic compounds, we begin with the scents emitted by the blossoms of ancient plants. Magnolias (*Magnolia* sp.), water lilies (*Nuphar* sp.), and sweet shrub or spicebush (*Calycanthus occidentalis*) (Figure 2.6) developed early in evolution among the flowering plants (angiosperms). Their flowers tend to give off faint smells of hydrocarbons, sometimes with a fermented note. Their floral structures, with petals open during the day, have relatively simple, sturdy shapes, with petals arranged in spirals. These are typical features of ancient floral structures. They are mostly pollinated by beetles. During evolution, beetles were the first pollinators,

Figure 2.6 Ancient flowers that are beetle-pollinated: (a) Yulan magnolia (*Magnolia denudata*). (b) Water lily (*Nuphar* sp.). (Photo by Verena Rau.) (c) Sweet shrub or spicebush (*Calycanthus occidentalis*).

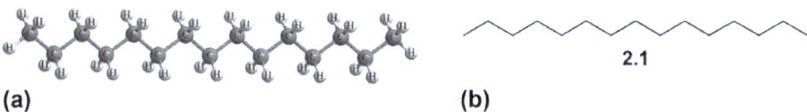

(a)   **(b)**

Figure 2.7 Molecular structure (a) and line structure (b) of a molecule of pentadecane. Pentadecane **2.1**, with the molecular formula of $C_{15}H_{32}$, is a volatile organic compound emitted by some magnolia flowers.

perhaps inadvertently so, by feeding on the flower petals and in the process transporting pollen onto the flower stigmas.

The flowers of the yulan magnolia (*Magnolia denudata*, Figure 2.6(a)), a species originally from China, have a faint smell sometimes described as waxy. Their volatile oils contain a high percentage (about 60%) of the hydrocarbon pentadecane, $C_{15}H_{32}$, with fifteen carbon atoms in a straight-chain, *i.e.* unbranched, sequence.[7] Figure 2.7 shows the complete molecular structure as well as the much simpler line structure of pentadecane **2.1**. Line structures will be used throughout this book. The shown structures present a very common shape or *conformation* of a molecule of pentadecane. We need to remember that single bonds between carbon atoms, as seen in pentadecane, are flexible, and allow atoms around them to be rotated, leading to a dynamic mixture of molecular shapes of the same compound. On the other hand, the carbon–carbon double bonds that we'll encounter in later molecules are rigid and do not allow free rotation of atoms bonded to them. For descriptions on how to understand line structures you may want to refer to the Glossary at the end of this book.

2.5   Volatile Alcohols, Aldehydes, and Esters

Functional groups, like alcohol (OH), aldehyde (CHO), or ester groups (COOR) can be attached to the hydrocarbon structures in plant volatiles, replacing hydrogen atoms in various positions in the molecules (Figure 2.8). Simple alcohols, aldehydes, and esters are commonly part of sweet floral fragrances that are attractive to bees and butterflies, and sometimes to moths as well. As an aside, these groups of compounds are generally perceived as pleasant by humans. Octanal **2.2** is an aldehyde with a fruit-like odor that occurs naturally in citrus oils. Esters are present in many floral scents, also in ripe fruits, like ethyl 2-methylbutyrate **2.3** in ripe apples, making them attractive to feeding animals, including fruit flies. Compounds that have a benzene ring in their structure, like 2-phenylethanol **2.4** and benzyl acetate **2.5** (Figure 2.8), are generally known as '*aromatic compounds*' in chemistry. The aromatic volatiles shown in Figure 2.8 are also aromatic in the sense of being pleasant-smelling compounds. 2-Phenylethanol is a common fragrant compound in floral scents, like rose essential oils (see Table 2.1). Benzyl acetate is an ester that represents a major fragrance component in jasmine flowers (Figure 2.3(b)). Molecules without aromatic rings, like octanal **2.2** or 2-methylbutyrate **2.3**, are known as '*aliphatic*'.

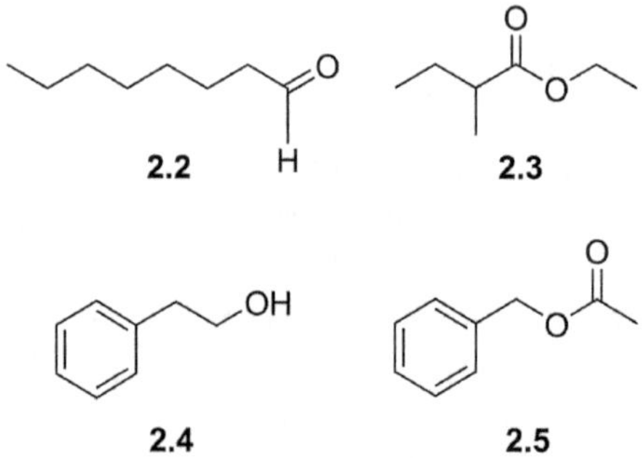

Figure 2.8   Aliphatic and aromatic volatiles with pleasant scents. Octanal **2.2** is an aliphatic aldehyde with a fruit-like odor. Ethyl 2-methylbutyrate **2.3** is an ester in ripe apples. The aromatic compound 2-phenylethanol **2.4** has a floral scent. Benzyl acetate **2.5** is the major fragrance component of jasmine flowers.

2.6 A Wealth of Terpenes

By far the largest group of plant volatiles comprises the family of *isoprenoids* or *terpenes*.[8] They are found in flower scents in various concentrations and are also part of defensive mixtures in leaf oils, as will be shown later. The name 'terpene' is derived from 'oil of turpentine', obtained from pine trees. Oil of turpentine is a mixture of terpenes with a characteristic odor. Terpene molecules are composed of five carbon units, the *isoprene units* **2.6** (Figure 2.9). Therefore, the total numbers of carbon atoms in terpene structures are multiples of five. The majority of volatile terpenes belong to the class of *monoterpenes*, with ten carbons, thus are composed of two isoprene units. The unique *isoprenoid* structure is the result of specific biosynthetic pathways that lead from carbohydrates, obtained from photosynthesis, to terpenes. A smaller group of volatile terpenes comprises the *sesquiterpenoids*, with fifteen carbons each in their molecular structures, the result of three isoprene units. These molecules are still small enough to evaporate at elevated temperatures. Larger terpene molecules, composed of four or more isoprenoid units, are not volatile at ambient temperatures.

Figure 2.9 shows examples of common monoterpenes and sesquiterpenes that occur in various percentages in many floral scents. (Note that several are listed in Table 2.1 as part of rose oil.) Their common names often recall plants that they can be found in. Geraniol **2.7** occurs in scented geraniums. (There are also systematic names of these compounds, geraniol being (*E*)-3,7-dimethyl-2,6-octadien-1-ol. Common names are easier!) The smells of monoterpenes are generally perceived as pleasant by humans and are attractive to insects like butterflies, moths, and bees. Functional groups like alcohol groups or aldehyde groups further contribute to sweet odor notes.

The examples of terpene structures shown in Figure 2.9 also give us a chance to look at different types of *isomers*, *i.e.* molecules that have the same *molecular formulas*, with the same number and same kinds of atoms, but with different structures. This results in molecules of different compounds and different properties. The phenomenon of isomerism adds to the enormous variety of potential organic structures.

Depending on the season and the cultivar, essential oils from the damask rose (*Rosa damascena*, Figure 2.3(a)), a major source of rose oil, can contain around 20% of sweet-smelling geraniol **2.7**, $C_{10}H_{18}O$

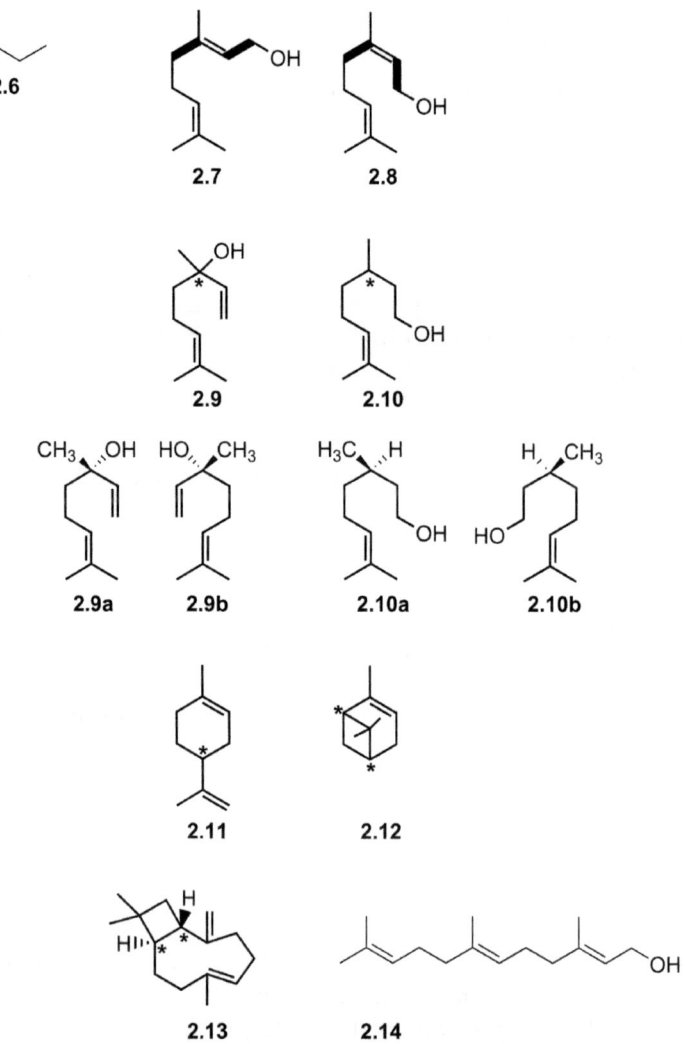

Figure 2.9 Common terpenes in floral scents and a discussion of isomers. The isoprene unit **2.6** is found in all terpenes. The sweet-smelling monoterpenes geraniol **2.7** and nerol **2.8**, with slightly different floral scents, are *trans/cis* isomers. Linalool **2.9**, citro-nellol **2.10**, limonene **2.11**, and α-pinene **2.12** are examples of monoterpenes with chiral centers. The two enantiomeric struc-tures of linalool **2.9a, b** (specifically named (*R*)-(−)-linalool and (*S*)-(+)-linalool), and of citronellol **2.10a, b**, are shown in more detail. Caryophyllene **2.13** and farnesol **2.14** are examples of sesquiterpenes. Chiral centers are indicated with asterisks (*).

(Figure 2.9). Nerol **2.8**, also with the molecular formula of $C_{10}H_{18}O$, has a similar, but not identical structure. We need to remember that double bonds induce rigidity in a molecule since there is no free rotation for atoms bonded to them. Therefore, if we focus on the single bonds shown as bold lines adjacent to the upper double bonds in **2.7** and **2.8**, we note that they appear at opposite sides in geraniol, whereas in nerol they are on the same side of the double bond. Such a relationship is known as *cis/trans isomerism*, and nerol is the *cis* isomer of geraniol. These are slight structural differences only, but the scent of nerol is perceived as somewhat different by humans (and probably also by insects). Nerol is found in rose oil as well, but in much smaller quantities than geraniol.

Linalool **2.9** also has the molecular formula of $C_{10}H_{18}O$, therefore is an isomer of geraniol and nerol. But its structure is altogether different, with the OH group and double bond located at different positions than in the molecules of geraniol and nerol. Isomers that have such different connectivities of their atoms are known as *constitutional isomers*. Heavy, sweet scents from linalool attract moths, like sphinx moths (Figure 2.10(a)).

With linalool we encounter an additional type of isomerism, namely molecules that are either left-handed or right-handed, *i.e.* they are mirror images of each other that are non-superimposable. The mirror images are known as *enantiomers*. This type of isomerism occurs if molecules have centers of asymmetry, known as *chiral centers*, marked with * in the molecular structures in Figure 2.9. Chiral centers in molecules can be identified as carbon atoms having

(a) **(b)** **(c)**

Figure 2.10 Volatiles from flowers attract insects. (a) A white-lined sphinx moth (*Hyles lineata*) is attracted by the volatiles of a desert broomrape (*Orobanche cooperi*). (b) Flowers of carnations emit sweet-smelling caryophyllene. (c) Fragrant cyclamen flowers emanate the sesquiterpene farnesol.

four different atoms or groups attached to them. This feature causes a molecule and its mirror image to have non-identical three-dimensional structures. Many monoterpene molecules are asymmetric and have one or more carbon atoms that are chiral centers. The two enantiomers of linalool, **2.9a** and **2.9b**, shown with bold and dashed bonds attached to the chiral centers to represent the three-dimensional aspect of the bond arrangements, have different smells. The scent of (*R*)-(−)-linalool is described as woody floral. It provides its woody lavender scent to lavender (*Lavendula* spp.) and is the major odor component of lavender herbs, with about 40% of linalool in its essential oils, where they probably have more of a defensive role. Its enantiomer, (*S*)-(+)-linalool, has a scent described as sweet floral and is found *e.g.* in the herb coriander or cilantro. Both enantiomers of linalool occur in rose oil. In general, plants may contain one of the enantiomers exclusively or a *racemic mixture*, *i.e.* a 1:1 mixture of both enantiomers. Rose oil also contains about 40% of citronellol **2.10**. Its two enantiomers **2.10a** (found in citronella oil) and **2.10b** (occurring in rose oil, also in pelargonium geraniums) are shown in Figure 2.9. Note that citronellol, $C_{10}H_{20}O$, is quite similar in structure to geraniol and nerol, but has only one double bond. Compare also with Table 2.1 for %-content and aroma notes of the above-mentioned monoterpenes.

The three-dimensional shapes of molecules and how they interact with the sensors of smell in insects or other animals (including humans) affect the perception of odor of the volatiles. Molecular asymmetries often influence how we perceive the smells of the respective compounds. Insects most likely experience them as different, too.

Many volatile terpenes are very common and are found in various plant scents, but in different concentrations. Limonene **2.11**, with the smell of citrus, is a frequent component of floral smells, especially in citrus plants, and is also found as traces in rose scents. Note that the molecule has a chiral center, one of the enantiomers having the smell of lemon and the other of orange. The smells are related but different nevertheless. The widespread α-pinene **2.12** is a major component of pine oil as its name suggests, and it contributes 40–50% to the essential oils of rosemary herb, acting more as a plant defense in both plants. Small amounts of α-pinene can be found in the essential oils of rose blossoms as well.

Examples of two common sesquiterpenes with pleasant scents are caryophyllene, $C_{15}H_{24}$, **2.13**, found in carnations (Figure 2.10(b)) and in cloves, and farnesol, $C_{15}H_{26}O$, **2.14**, occurring in cyclamens (Figure 2.10(c)) and roses.

2.7 Sulfur- and Nitrogen-containing Volatiles That Attract Flies

Volatile organic compounds from flowers that contain sulfur or nitrogen atoms in their molecular structures contribute to plant odors that humans experience as unpleasant, but that are inviting to flies, some beetles, or to small insects called gnats. Distinct floral pigments enhance the attraction towards these pollinators. Large clusters of small, off-white flowers, as in angelica plants (*Angelica* sp.) (Figure 2.11(a)) or cow parsnip (*Heracleum* sp.), together with off-odors, attract flies and some beetles. Brown or maroon pigments that may simulate decomposing animal materials often combine with foul smells that attract gnat and fly pollinators.

Some of these foul-smelling compounds are shown in Figure 2.12. The simple aliphatic amine 1-aminohexane **2.15** contributes to the off-odor of cow parsnip flowers (*Heracleum lanatum*) and related plants. Volatile amines have odors that are generally highly unpleasant to humans. The compound indole **2.16**, chemically an aromatic amine, is part of the complex mixture that contributes to the strong fragrance of hyacinths (Figure 2.11(b)), a floral scent that we generally consider as pleasant. Surprisingly, indole is also found in human feces and contributes to the fecal odor there. But when

(a)   (b)   (c)

Figure 2.11   Flowers that emit volatiles with unpleasant odors. (a) Inflorescence of angelica (*Angelica* sp.) with fly pollinator. (b) Hyacinths (*Hyacynthus*). (c) Flower of a *Stapelia gigantea* or carrion plant. (Photo: Gilbert8888. Wikimedia Commons. https://upload. wikimedia.org/wikipedia/commons/d/d7/Aasblume_Aug_2005. jpg, (accessed September 2016).)

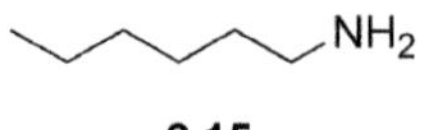

Figure 2.12 Examples of floral volatiles with foul smells. The amines 1-aminohexane **2.15** and indole **2.16**, and the sulfurous dimethyl disulfide **2.17** and dimethyl trisulfide **2.18** have highly unpleasant odors (to humans). So does hexanoic acid (caproic acid) **2.19**, a simple fatty acid.

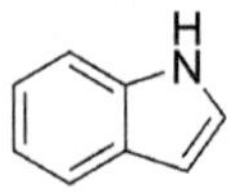

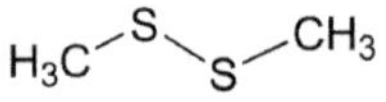

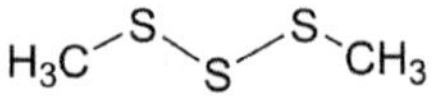

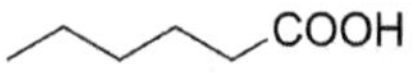

dissolved in a mixture at very low concentrations indole exhibits a floral smell.

Sulfides and disulfides, like dimethyl disulfide **2.17** or even dimethyl trisulfide **2.18**, with three sulfur atoms bonded to each other, are part of the scents in the several types of plants appropriately named corpse or carrion flowers (*Rafflesia* or *Amorphophallus* spp., Figure 2.3(c)).[9] The foul odors emitted by the brownish-purple flowers of carrion plant (*Stapelia gigantea*, Figure 2.11(c)), contain sulfides and indole. Their volatile mixtures have also been found to contain simple *fatty acids, i.e.* hydrocarbon chains with a carboxylic acid group attached, like (mal)odorous hexanoic acid (also known as caproic acid) **2.19**.[10] Flies are attracted by inflorescences that emit these fetid smells and accomplish pollinating the flowers while visiting. Distinct patterns, as in the flowers of *Stapelia*, further guide the insects to the pollen.

2.8   Sweet, Nutritious Nectar in Flowers

Many plants reward pollinating insects with offerings of sweet nectar in their flowers. These sugary solutions are stored in *nectaries*, which tend to be shiny, yellowish-green surfaces in flowers, sometimes with droplets of nectar visible. Flower nectars are a primary source of nutrition for butterflies and are attractive to bees, wasps, and some beetles as well. Quantities of nectar in a flower are small, just enough

Figure 2.13   Nectars in flowers. (a) Showy nectar guides on the petals of a purple Douglas iris (*Iris douglasiana*). (b) Desert rock nettle (*Eucnide urens*) from SW USA, with light-yellow nectaries visible, with beetles. (c) Flowers of yellow jessamine or Carolina jasmine (*Gelsemium sempervirens*) contain toxic nectar.

to be inviting to pollinators, but not enough to satiate a visiting insect. In other words, pollinators have to visit several flowers in order to get fed, which again benefits the plants. Offering nutritious nectar (and pollen, as addressed later) to insects in order to obtain the services of pollination illustrates examples of *mutualism*, *i.e.* mechanisms that mutually benefit plants and insects. Figure 2.13 shows examples of flowers that provide nectar to their pollinators.

In some plants, nectar is also produced outside flowers, in so-called *extrafloral nectaries*, *e.g.* on plant stems or tree trunks, attracting ants that guard against herbivores, or on leaves converted to traps in carnivorous plants. These mechanisms will be further elaborated in later chapters.

Nectars are essentially mixtures of water with sugars dissolved in them in various concentrations and proportions. Figure 2.14 shows examples of compounds found in plant nectars. The sugars, or carbohydrates, are the monosaccharides D-glucose **2.20** and D-fructose **2.21**, and the disaccharide sucrose **2.22**. Their cyclic structures are shown as Haworth projections, which are commonly-used, simplified presentations of the three-dimensional perspectives of the molecules. D-glucose is shown in its α-form, pointing to the specific three-dimensional arrangement at C1 in the molecule. Note that the molecules have numerous chiral carbon atoms. (They are marked with an asterisk (*) in glucose **2.20**.) The molecular structures of sugars characteristically have many OH groups, making them highly polar compounds and therefore very soluble in the aqueous nectar solutions. The concentrations of the sugars in the nectar solutions vary from about 10% to 75%, depending on the plant and the developmental stage of the flowers.

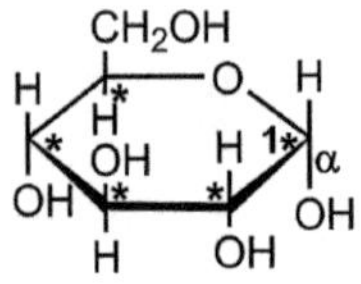

Figure 2.14   Nectar components. The monosaccharides D-glucose **2.20** (α-form) and D-fructose **2.21**, and the disaccharide sucrose **2.22** and minor concentrations of amino acids, like phenyl-alanine **2.23** and proline **2.24**, are commonly found in nectar solutions. Floral nectars can also contain toxins, like the alkaloid gelsemine **2.25** in wild jessamine (*Gelsemium sempervirens*).

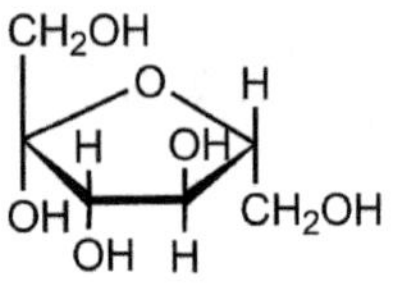

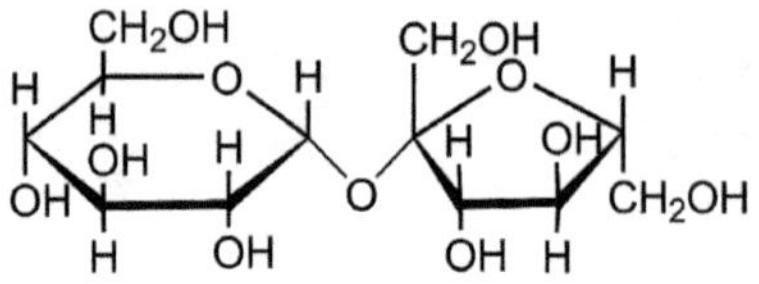

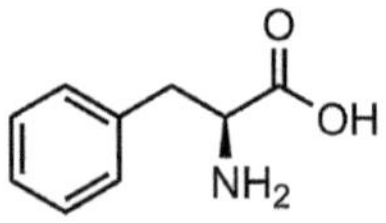

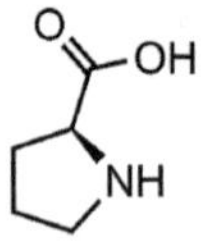

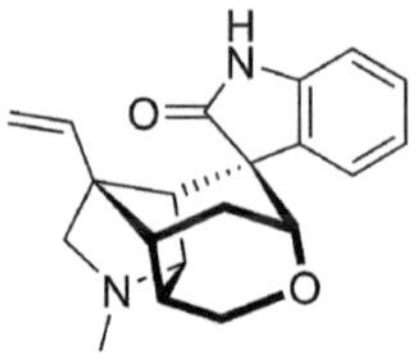

Low, but measurable concentrations of amino acids, like phenyl-alanine **2.23** or proline **2.24** (Figure 2.14), are dissolved in the nectars as well. As butterflies feed mostly and often exclusively, on flower nectars, the amino acids in nectars are important sources of nitrogen-containing primary metabolites for the insects. Some carrion and dung flies are especially attracted to flowers with nectars that have higher concentrations of amino acids and feed on their nectars.

Floral nectars are also an important source of water for insects, especially in desert areas.

Interestingly, quite a few flower nectars contain small amounts of toxins, *i.e.* compounds that can harm or even kill pollinators. Rhododendrons and related plants are well-known for their toxic floral nectars that contain grayanotoxins, a group of terpenes with twenty carbon atoms (a multiple of five!) in their complex molecular structures. While bees apparently can tolerate a limited amount of the nectar poisons, honey produced by bees visiting rhododendrons is toxic as well. (More on this topic will be addressed in the later chapter on honey). Yellow jessamine or Carolina jasmine (*Gelsemium sempervirens*) (Figure 2.13(c)), an evergreen vine originally from southwestern North America, contains the alkaloid gelsemine **2.25**, a compound with a complex molecular structure. *Alkaloids* are generally secondary metabolites that contain nitrogen in their molecular structures; we will encounter them especially among the chemical plant defenses. Gelsemine is toxic to bees as well as to humans. Alkaloid toxins tend to affect the taste of the nectars too, adding a bitter note. This seems counterproductive from the plants' point-of-view, and various theories have been developed about the presence of poisonous, distasteful compounds in nectars. Defensive effects towards certain insects, like ants, that might disturb pollinators, or plants' selectivity towards certain pollinators that can tolerate low doses of the toxins have been suggested. A somewhat bitter taste of the nectar may limit the amount ingested and may lead to shorter, but more repeated visits. On the other hand, some floral nectars that contain the alkaloids caffeine or nicotine were found to be preferred by honey bees. Studies on nectar toxins and their selective effects on insects are fields of active research.[11]

2.9 Plant Colors and Nectar Guides

Bright colors of flowers visible to potential pollinators combine with floral scents and nectar rewards to attract pollinating insects. Examples of flowers of many shades and colors have been shown in Figures 2.2(a) and (b), 2.3, 2.6, 2.10, 2.11 and 2.13. Flower colors are a type of directional communication with potential pollinators, *i.e.* they must be seen by approaching insects. The colors provide shorter distance communication to insects than smells, but are generally longer-lasting than floral scents.

Colors in plants are created by *pigments*, *i.e.* by compounds that absorb a section of the electromagnetic spectrum of sunlight and reflect or transmit the remaining wavelengths (Figure 2.15). There are also structural colors in plants, like the hues that are created by fine hairs, or by spines that reflect sunlight, but they are of minor importance for producing colors in plants. (This is quite unlike colors of insects, as will be shown in Chapter 5.3.) Many insects, including honey bees and bumble bees, perceive wavelengths of around 360 nm which is near-UV light (invisible to humans). Bees see blues and yellows especially well, but do not recognize reds. Butterflies are

Figure 2.15 Colorful pigments in flowers. (a) Eupatory (*Ageratina* sp.), a member of the Aster family, has purple inflorescences with anthocyanin pigments that are attractive to butterflies. (b) The inflorescence of the food plant amaranth (*Amaranth* sp.) gets its purple color from betalain pigments. (c) The carotenoids in petals and stamens of California poppy (*Eschscholzia californica*) are clearly visible to bumble bees.

attracted by a wide spectrum of colors, including near-UV. Carrion flies that feed on plant nectars besides carrion have a preference for brown and purple flowers.

Let us focus on typical molecular structures of plant pigments. This will also provide a general introduction to the common structural features of organic molecules that absorb light within the range of spectral light visible to humans – or insects. When a pigment absorbs a section of the electromagnetic spectrum of sunlight it reflects or transmits the non-absorbed portion. If absorption occurs in the part of the spectrum visible to humans, then we see the reflected or transmitted part and a colorful pigment. Similarly, if absorbed wavelengths occur within a section of sunlight seen by an insect, the corresponding reflected or transmitted wavelengths are perceived by the insect. The structure of a molecule and its pattern of chemical bonds can convey whether a compound appears colored to us or not. Upon examining Figures 2.16, 2.17 and 2.18, note that all the shown molecular structures of pigments feature long sequences of double bonds alternating with single bonds, the so-called *conjugated double bonds*. This is a characteristic of organic molecules that strongly absorb light. The longer the sequence of conjugated double bonds, the longer is the wavelength (and the lower is the energy) of absorbed light, *i.e.* the absorbed wavelengths move towards the red. A pigment that absorbs mostly short-waved, higher energy (blue) light would appear to us as yellow or orange, whereas one that absorbs longer

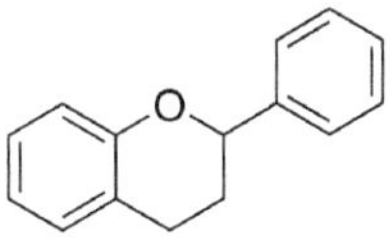

Figure 2.16 Flavonoid pigments. The common flavonoid structure **2.26** is found in flavones like luteolin **2.27**, as well as in anthocyanidins like pelargonidin **2.28** and its glycoside pelargonin **2.29**. The anthocyanin pelargonin has two glucose groups ("gluc") bonded to its molecules.

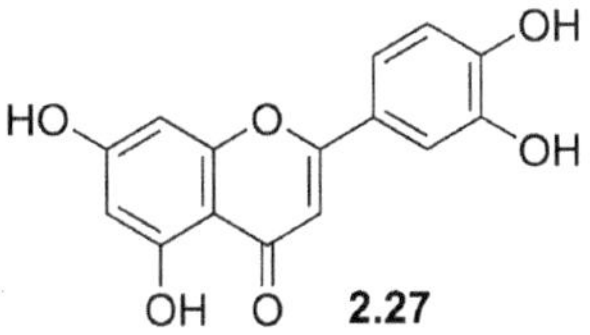

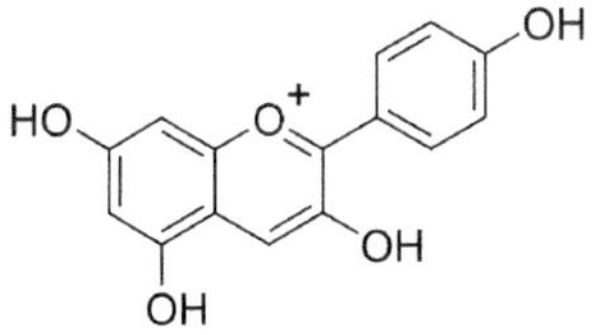

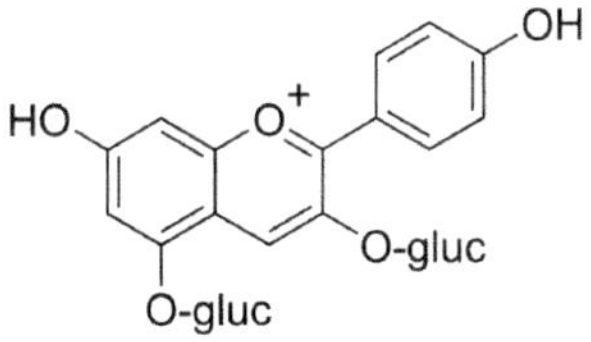

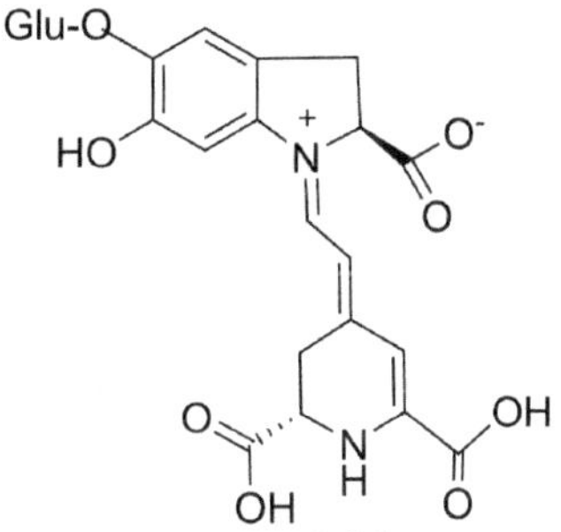

Figure 2.17 Betalain pigments. The red pigment betanin **2.30**, found in red beets (goosefoot family), is an alkaloid.

wavelengths would look blue or green to humans. Sequences of just a few conjugated double bonds lead to absorption in the UV. The corresponding reflected wavelengths are invisible to humans, but are perceived by many types of insects.

There are only a few chemical families of plant pigments. We first focus on pigments that are dissolved in the aqueous cell saps. The vacuoles in plant cells are often colored by them. Molecules of these pigments are commonly bonded to sugar molecules like glucose. This property further enhances the pigments' solubility in water due to the increased number of polar functional groups, such as the OH groups,

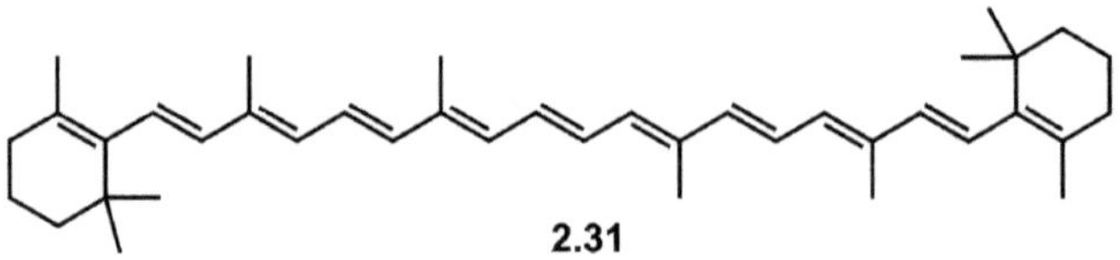

Figure 2.18   Carotenes and xanthophylls. β-Carotene **2.31** and the xanthophyll lutein **2.32**.

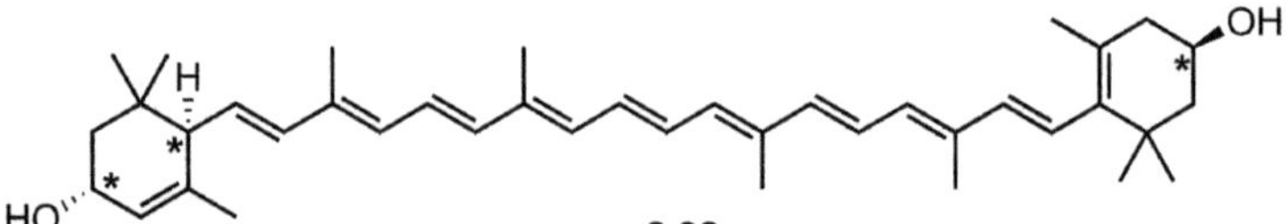

from the connected sugar molecule. Water-soluble pigments can easily be transported in the aqueous plant saps. Molecules consisting of a carbohydrate part bonded to a non-carbohydrate moiety (the *aglycon*) are generally known as *glycosides*.

Most water-soluble plant pigments belong to the family of *flavonoids* (Figure 2.16), providing a wide palette of plant colors, from off-white to yellows to purples and blues. The common molecular structure **2.26** that is found in all flavonoids, together with examples of typical pigments, is shown in Figure 2.16. Pigments of the sub-family of *flavones* and *flavonols* provide off-white and yellow colors as well as strong UV-absorption, properties that are visible to bees, moths, and butterflies (see Figure 2.19). An example of a common yellow flavone pigment is luteolin **2.27** occurring, *e.g.*, in off-white and yellow chrysanthemum flowers. *Anthocyanins* are another most common subfamily of flavonoid pigments. They provide pink, purple, and blue colors to plants. The pigment molecules are bonded to sugar molecules like glucose, making them glycosides. This increases their solubility and transportability in the aqueous plant system. Figure 2.16 shows pelargonidin **2.28**, which is the aglycon of the anthocyanin pelargonin **2.29**. Both are reddish pigments. Flowers colored by anthocyanin pigments, with their blue to purple colors, greatly attract bees and bumble bees.

Anthocyanin pigments frequently form bonds with other flavonoid molecules, leading to lengthened systems of conjugated double bonds in their molecules, and consequently to more intense colors and a shift to absorption of longer wavelengths of light. Anthocyanins also bond to metal ions that plants obtain from the soil, resulting again in variations of their colors. Changes in pH affect anthocyanin

(a)   (b)   (c)

Figure 2.19   UV-absorbing nectar guides. (a) A cluster of monkeyflowers (*Mimulus guttatus*) in sunlight. (Photo by Glenn Keator.) (b) A single monkeyflower in daylight and (c) under UV-light. (Photos by Megan Lynn Peterson.)

colors as well. All these various conditions lead to a lot of color variations among anthocyanin pigments.

A much smaller family of pigments that also provides purple to purplish-red colors to plants is the family of *betalains* (Figure 2.17). These water-soluble pigments are only found in plants of certain plant families, all in the order Caryophyllales, like the cacti (Cactaceae) or the goosefoot family (Chenopodiaceae) or the amaranth family (Amaranthaceae), as represented with the example of the food plant amaranth (*Amaranthus* sp.) shown in Figure 2.15(b). Interestingly, the purplish-reds in plants of these families are exclusively created by betalains. Betanin **2.30** is the well-studied red betalain pigment from red beets (goosefoot family). When comparing its molecular structure to anthocyanin structures (Figure 2.16) it is easy to note the major structural differences in betalain pigments, including the presence of nitrogen atoms. Betalains are alkaloids which absorb and reflect wavelengths similar to anthocyanin structures. Note again the long sequences of conjugated double bonds that lead to strong absorption of light. The color of *Ageratina* in Figure 2.15(a) is quite similar to the color of the inflorescence of amaranth (Amaranthaceae) in Figure 2.15(b). Only a chemical analysis would show up the substantial differences in the structures of their molecules. Betalains are chemically much more stable than anthocyanins, *e.g.* when exposed to changes in pH, and do not change color easily. (In fact, they are so stable that red beet juice is used as a food additive, *e.g.* to give yoghurt or ice cream a pink color.) Betalain pigments similar in structure to betanin **2.30** are found in the petals of pink cactus blossoms (Figure 2.2(a)).

Hydrophobic or lipophilic pigments are stored in special compartments in plant cells, called *plastids*. Most famously, the photosynthetic pigment chlorophyll a and related chlorophylls, that act as accessory pigments, are stored in chloroplasts. *Carotenes* and the related *xanthophylls* are generally known as *carotenoids* (Figure 2.18). They provide the deep-yellow, orange, and orange-red pigments in flowers like California poppies (Figure 2.15(c)). They are also responsible for the yellow pigmentation in many fruits and in leaves. Carotenoids are terpenes, with forty carbon atoms (*i.e.* assembled from eight isoprene units) in their molecular structures. They are mostly hydrocarbons, thus water-insoluble. Note the long sequence of conjugated double bonds in β-carotene **2.31**, the yellow to orange pigment in many flower petals (and in carrot roots, as well). Xanthophylls are carotenoids that include an oxygen function, like an OH group or two, in their structures. Lutein **2.32** is an example of a most common xanthophyll, found *e.g.* in the petals of common garden flowers like marigolds (*Tagetes* sp.) or yellow nasturtiums (*Tropaeolum* sp.). Note that lutein has several chiral centers, marked with asterisks in Figure 2.18. Xanthophylls are also found in green leaves and many fruits. Carotenes and xanthophylls absorb light in the UV as well, thus are clearly visible to many insects.

Flower petals can contain several different types of pigments. In addition, concentrations of the pigments are variable, leading to an almost endless palette of flower colors. Think of "black" tulips which actually contain a high concentration of purple anthocyanins. Petal colors change with age of the flowers, often fading due to the decomposition of pigments, leading to low attraction for visiting insects.

In connection with the earlier descriptions of floral nectars, pigment patterns called *nectar guides* are designs in the flowers that lead potential pollinating insects to the nectaries and with this to brushing pollen onto the stigmas in flowers. Nectar guides are sometimes visible to humans, as in Figure 2.13(a), but are often mostly UV absorbing, as illustrated in Figure 2.19, *i.e.* invisible to humans, yet visible to many insects. The flowers of monkeyflowers (*Mimulus guttatus*) (Figure 2.19(a) and (b)) appear as yellow to us, with various flavonoids and some carotenoids providing the colors. When the flowers are viewed under UV-light, the pigments show a dark pattern of ultraviolet absorption that guides insects to the nectar (Figure 2.19(c)).[12] Nectar guides and nectaries are commonly absent in beetle-pollinated plants.

Figure 2.20 Flowers that change color after pollination. Lantanas (*Lantana camara*) (a) before and (b) after pollination. (c) Lupine (*Lupinus* sp.) showing unpollinated blossoms at the top of the inflorescence and purple pollinated blossoms below.

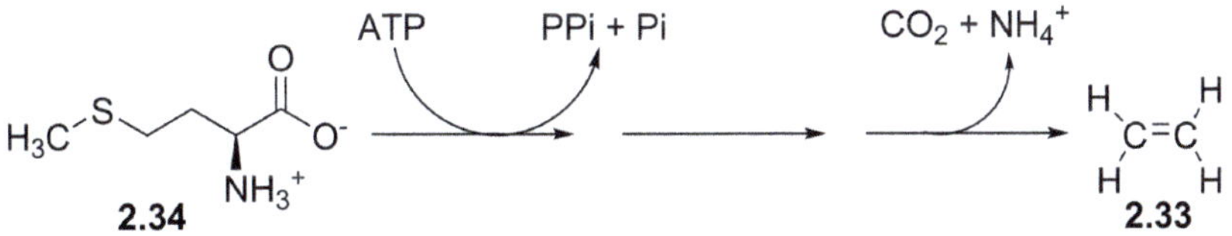

Figure 2.21 Biosynthesis of ethylene (or ethene) gas. Ethylene **2.33** is a plant hormone that forms in response to stress. It is also involved in plant growth. Ethylene is formed from the amino acid methionine **2.34** in a series of enzyme-catalyzed steps.

As a further guide to pollinating insects, several types of flowers change color after being pollinated (Figure 2.20). The flowers of lantanas (*Lantana camara*), a widespread, sometimes weedy plant originally from tropical Central and South America, change color from yellow (nonpollinated) to purplish-red (pollinated) (Figure 2.20(a) and (b)). The orange-yellow coloration is provided by the pigment β-carotene. As a response to pollination, a purple anthocyanin forms that changes the overall color to a deep red. In another example, purple lupines (*Lupinus* sp.), change the color of the upper lips or 'banners' of their flowers from white to purple, with pigments provided by anthocyanins (Figure 2.20(c)). Bees and bumble bees will seek out lupines with white banners.

How are such color changes induced? When flowers are successfully pollinated, pollen tubes grow into the stigma, basically destroying plant tissue. This process is likely mediated by ethylene (or ethene) gas **2.33**, a plant hormone that generally forms in response to stress. Figure 2.21 shows a series of reaction steps that take place during the *biosynthesis*, *i.e.* the biological synthesis, of ethylene that occurs in the living plant. Ethylene gas is biosynthesized in a sequence of steps from the amino acid methionine **2.34**. The reaction steps are powered by ATP (adenosine triphosphate) liberating PP_i (inorganic pyrophosphate) and P_i (inorganic phosphate) and catalyzed by specific biological catalysts or *enzymes*.[13] Ethylene gas is also formed during the ripening of fruits.

2.10   Pollen-rich Flowers for Bees

Pollen, formed in the anthers of many flowers, provides a major reward for visiting insects. Bees and bumble bees are especially attracted by masses of pollen, as shown in Figure 2.22(a) and (b). Some flowers, like poppies, offer pollen only (and no nectar) as a reward to visitors (Figures 2.15(c) and 2.22(b)). For the visiting insects pollen is food, providing them with nutritious proteins and starch. A single flower alone does not offer enough pollen to satiate an insect. Therefore, pollinators have to move from flower to flower, and while collecting pollen they also leave some pollen grains on the stigmas and thus pollinate the plants.

Pollen grains contain the male information of a plant in the form of *gametes,* or male reproductive cells. Pollination is the transfer of pollen from the anthers of a flower to the stigma, *i.e.* the female part, in another plant of the same species (Figure 2.1). If successful, this process induces fertilization, starts the production of seeds, and thus promotes the reproduction of the plant. Pollen grains

(a)   (b)   (c)

Figure 2.22   Pollen-rich flowers and pollinators. (a) Wild lilac (*Ceanothus* sp.) offers plenty of light-yellow pollen to a honey bee. (b) Native bees are attracted by the ample, deep-yellow stamens full of pollen in the flower of a Matilija poppy (*Romneya* sp.). (c) Forget-me-nots (*Myosotis* sp.) have very small pollen grains.

vary greatly in size. They can be as tiny as five micrometers in diameter, like the pollen grains produced by forget-me-nots (*Myosotis* sp., Figure 2.22(c)), or much larger, with a diameter of 100–200 micrometers, as found *e.g.* in plants of the cucumber family (Cucurbitaceae).[14]

Insect-pollinated plants must produce pollen that is designed to be appealing to insects, and flowers must have stigmas at the proper developmental stage and with shapes that facilitate the deposit of pollen. Pollen of a plant is species specific. Only stigmas of flowers of the same species are fertilized by it. The colors of pollen grains are commonly yellow, due to carotenoid pigments, but also purple or light blue. These colors are attractive to insects and visible to bees and bumble bees. Pollen grains have characteristic scents, too, in the form of volatiles, like terpenes, similar to those shown earlier as floral volatiles. Characteristic scents combined with colors and shapes of pollen grains are recognized by insects.

The rewards for insects performing the essential service of pollination are nutritious proteins, starch, and some lipids. The content of starch in pollen grains can be up to 22%. The protein content of pollen grains can vary from 2.5–61%. It is especially high in so-called 'buzz-pollinated' flowers, like bee-pollinated ones, where the rapid movement of the insect's wings shakes off pollen onto the insect.

The digestion of starch and proteins supplies insects with important nutrients. Starch and proteins consist of *polymers*, *i.e.* very large molecules with repeat patterns called *monomers*. Starch is a polymeric carbohydrate or *polysaccharide*. It consists of two related polymers, amylose and amylopectin. Glucose **2.20**, specifically D-glucose, is the monomer in both of them, being repeated thousands of times in specific three-dimensional arrangements. The D-glucose monomers in starch are connected by 1,4-α-glycosidic linkages. The numbers refer to the specific carbon atoms of each ring linked to their neighboring glucose units, while 'α' defines the three-dimensional arrangement of the link. Figure 2.23 shows a segment of the polymers **2.35** with the characteristic linkages of their D-glucose molecules. While amylose is composed of long, unbranched chains of D-glucose units, amylopectin is a highly branched polymer. Animals, like insects (humans, too), have the proper enzymes in their systems to break down amylose and amylopectin into glucose and other simple sugars and can make use of them as essential nutrients.

Proteins are polymers composed of amino acid monomers. Specific enzymes induce loss of water molecules between the monomers leading to sequences of amino acids of any length linked by *peptide*

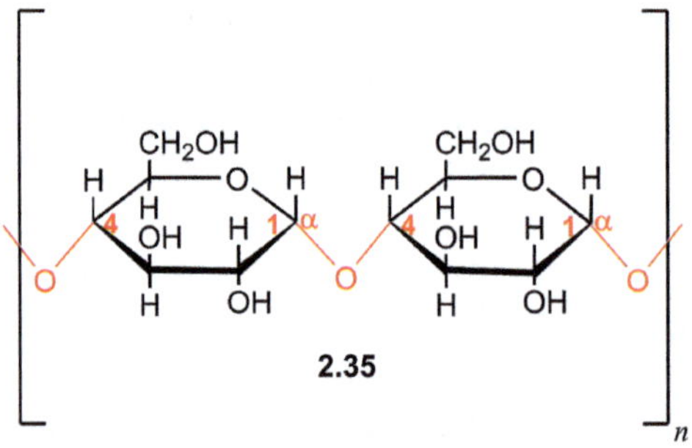

Figure 2.23   A sequence of D-glucose units in 1,4-α-glycosidic linkages.

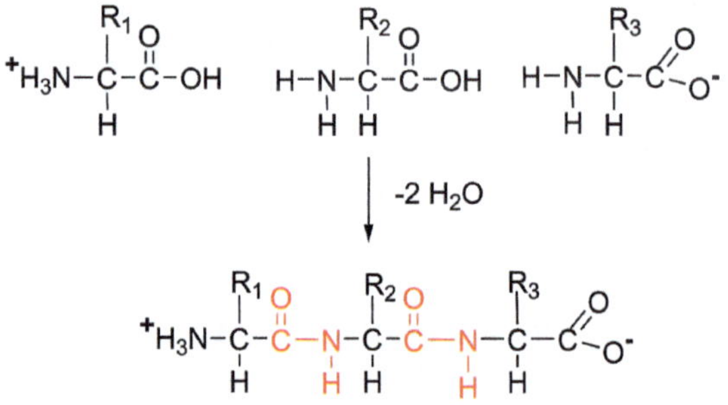

Figure 2.24   A sequence of three amino acid monomers linked by peptide bonds (shown in red) after loss of water.

bonds, as shown in Figure 2.24. Twenty different potential amino acid monomers, distinguished by varying attached groups (labeled as R_1, R_2, R_3 in Figure 2.24), can be combined in specific order and arrangement, with repetitions as well, to form a protein. The different protein molecules have defined three-dimensional shapes. This leads to a huge diversity of proteins, with a multitude of different functions.

The related peptides consist of smaller molecules composed of fifty or fewer amino acid monomers. An important type of function of proteins and peptides is as *enzymes* that catalyze the multiple steps in biochemical reactions in plant and animal systems. Enzymes are complex protein or peptide molecules, with three-dimensional shapes that fit the shapes of the biological molecules they are interacting with. Figure 2.24 shows a segment of an amino acid sequence **2.36** in a protein or peptide, the amino acid monomers being linked by peptide bonds shown in red.

Pollen also contains minerals, vitamins, and lipids, and therefore provides important nutrition for insects collecting it.

Pollen is well-known for its longevity. Pollen grains have a hard outer shell that contains 'sporopollenin', a polymeric substance with a structure and composition still widely unknown. Its chemical inertness makes it virtually impossible to dissolve or break down sporopollenin into smaller and simpler molecules that then could be analyzed. The longevity of pollen grains has made it possible to detect them in fossils. There is a vast diversity in pollen grains. They are species specific in shapes and in their chemical compositions. This specificity means that finds of pollen grains can lead to the determination of their plant origins.[15]

2.11 Orchid Strategies to Attract Insects

Orchids (Orchidaceae) are the largest and most diverse family of flowering plants. Members of this plant family have evolved – and keep evolving – not only complex forms of their flowers, but also a great diversity of pollination mechanisms. Charles Darwin, the father of evolution theory, was fascinated by the diversity of orchid structures. When critics of his earlier work on evolution doubted his theory of natural selection, he focused his research on orchids and their pollinators. He concluded that different orchid species must have evolved strategies that ensure cross-pollination, thus benefitting the species by producing greater numbers of viable seeds.[16]

Many types of orchids use amazing mechanisms to lure insects to pollinate their flowers. Elaborate flower shapes and intricate flower chemistry often deceive pollinators and can leave them without the reward of pollen. Several types of orchids attract potential pollinators with flower shapes that resemble particular female insects. Orchids of the genus *Ophrys*, found in Europe, Australia, South Africa, and South America are well-known for their intricate schemes to attract specific pollinators and then deceive them. The shapes of their flowers and their scents mimic female solitary bees. In the case of the bee orchid (*Ophrys apifera*, Figure 2.25), the flowers emit a scent mixture that closely resembles the pheromone of the female solitary bee *Andrena nigroaenea*. Note the patterning of the orchid flower. Male bees move in attempted copulations across the flower and in the process transport pollen, in the form of pollen packages called 'pollinia', onto the orchid stigma. The bees do not obtain pollen or nectar for food in the process, and the plants do not need to expend energy to produce extra pollen nor attractive nectar in their flowers.[17]

Figure 2.25   *Ophrys* orchids deceive bee pollinators. Flower of a bee orchid (*Ophrys apifera*) with distinct patterning that, combined with its scent, attracts male bees.
(Photo: Hans Hillewaert. Wikimedia Commons. https://upload. wikimedia.org/wikipedia/commons/e/ef/Ophrys_apifera_(flower). jpg (accessed September 2016).)

Pheromones are generally organic compounds or mixtures that are emitted by animals, usually in very small amounts, and that affect the behavior of other animals of the same species, either sexually or in alarm reactions or for trail following. The compounds that compose the pheromones given off by female *Andrena* solitary bees and that are also produced by *Ophrys* flowers have been analyzed by gas chromatography-mass spectrometry methods (Figure 2.5). They were found to be mixtures of several long-chain hydrocarbons. The pheromone components have odd numbers of carbon atoms, with 21–29 carbons bonded either with single bonds only or with a sterically-defined double bond in the hydrocarbon chain. Figure 2.26 shows the examples of tricosane **2.37**, a saturated (single-bonds only) hydrocarbon with twenty-three carbons, and (*Z*)-9-heptacosene **2.38**, a hydrocarbon with twenty-seven carbon atoms and one double bond in *cis* or (*Z*) arrangement. The all single-bonded chain in **2.37** has great flexibility within the molecule. Note that the rigid double bond in **2.38** leads to a bend in the structure. These hydrocarbons are waxy substances occurring on the plants' surfaces and not only attract pollinators, but also help prevent water-loss from the plants. The same or very similar compounds are found

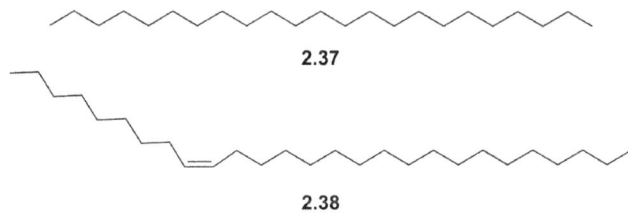

2.37

2.38

Figure 2.26   Examples of long-chain hydrocarbons **2.37** and **2.38** found in *Ophrys* flowers that mimic female bee pheromones.

in the pheromones of the targeted female bees. The compounds have a faint smell that excites the male bees.[18]

On a biochemical note: The odd numbers of carbon atoms in the long-chain hydrocarbons are the result of biosynthetic steps that lead to their formation in the plants. The hydrocarbons are derived from *fatty acids*, *i.e.* long-chain hydrocarbons with a carboxylic acid (COOH) functional group. Fatty acids in plants are biosynthetically composed from several acetate units, with two carbon atoms each, and therefore have even-numbered carbon chains, *e.g.* C_{12}, C_{14}, C_{16}, *etc.* Loss of carbon dioxide, CO_2, from the carboxylic acid molecules in a following step, results in the loss of one carbon atom per molecule, leading to hydrocarbons with odd numbers of carbon atoms.

2.12   Plants Attracting Insects and Coevolution

Offerings of nectar and pollen to obtain the services of pollination involved examples of true mutualism between plants and insects. Through evolution plants have developed – and keep evolving – elaborate mechanisms to attract suitable pollinators, even by sometimes deceiving them; in return, insects have evolved their own mechanisms to obtain the desired nutrition from flowers. Throughout this book we'll encounter examples of *coevolution* between plants and insects.

In 1964, P. R. Ehrlich and P. H. Raven published an article in which they emphasized the reciprocal aspects of many types of interactions between organisms of different species. If the organisms, like plants and insects, interact closely with each other, they are likely to exert a strong selective force on each other. This paper stimulated modern research on coevolution.[19]

2.13 Conclusions

Plants have evolved many chemical schemes to attract insects to pollinate their flowers and thus to promote reproduction of the plants. A combination of scents and pigments, as well as food offerings in the form of nectar and pollen, draw insects to visit. Components of plant scents or pigments, even if undetected by our nose or invisible to our naked eyes, have crucial and often very noticeable roles as plant attractants.

Plant compounds that invite insects to their flowers have provided an introduction to organic molecules. Molecular structures explain and allow us to predict many properties of compounds, like their volatility or their solubility in water or fats. The colorful appearance of pigments is related to the chemical structures of their molecules as well. Starch and protein molecules served as an introduction to polymers. Different types of isomers lead to a vast diversity of organic compounds.

In the following chapter we shall focus on plants that lure insects not only for pollination, but also as nutritional supplements for the plants.

References

1. L. M. Schoonhoven, J. J. A. van Loop and M. Dicke, *Insect-Plant Biology*, Oxford University Press, Oxford, 2005.
2. J. T. Knudsen, R. Eriksson, J. Gershenzon and B. Stähl, Diversity and distribution of floral scent, *Bot. Rev.*, 2006, 72, 1.
3. H. M. Schaefer and G. D. Ruxton, *Plant-Animal Communication*, Oxford University Press, Oxford, 2011.
4. P. Jolivet, *Interrelationships Between Insects and Plants*, CRC Press, Boca Raton, 1998.
5. M. Séquin, Volatiles for the Perfume Industry, in *Encyclopedia of Applied Plant Sciences*, ed. B. Thomas, B. G. Murray and D. J. Murphy, Academic Press, Waltham, MA, 2nd edn, 2017, pp. 393–398.
6. M. Séquin, *The Chemistry of Plants: Perfumes, Pigments, and Poisons*, Royal Society of Chemistry, Cambridge, UK, 2012, ch. 3, pp. 79–92.
7. H. Azuma, M. Toyota, Y. Asakawa, R. Yamaoka, J. G. Garcia-Franco, G. Dieringer, L. B. Thien and S. Kawano, Chemical

Divergence in Floral Scents of *Magnolia* and Allied Genera (Magnoliaceae), *Plant Species Biol.*, 1997, **12**(2–3), 69.

8. E. Breitmaier, *Terpenes: Flavors, Fragrances, Pharmaca, Pheromones*, Wiley-VCH Verlag, Weinheim, 2006.
9. G. C. Kite and W. L. A. Hetterschieid, Inflorescence Odours of *Amorphophallus* and *Pseudodracontium* (Araceae), *Phytochemistry*, 1997, **46**(1), 71.
10. A. Jürgens, S. Dötterl and U. Meve, The chemical nature of fetid floral odours in stapeliads, *New Phytol.*, 2006, **172**, 452.
11. J. D. Thomson, M. A. Draguleasa and M. G. Tan, Flowers with caffeinated nectar receive more pollination, *Arthropod-Plant Interact.*, 2015, **9**, 1.
12. M. L. Peterson, T. J. Miller and K. M. Kay, An Ultraviolet Floral Polymorphism Associated With Life History Drives Pollinator Discrimination in *Mimulus guttatus*, *A. J. Bot.*, 2015, **102**(3), 396.
13. C. Bowsher, M. Sterr and M. A. Tobin, *Plant Biochemistry*, Garland Science, New York, NY, 2008.
14. P. Willmer, *Pollination and Floral Ecology*, Princeton University Press, Princeton and Oxford, 2011.
15. P. H. Raven, R. F. Evert and S. E. Eichhorn, *Biology of Plants*, W. H. Freeman, New York, NY, 7th edn, 2005.
16. C. Darwin, *On the Origin of Species*, John Murray, London, 1859.
17. F. P. Schiestl, M. Ayasse, H. F. Paulus, C. Löfstedt, B. S. Hansson, F. Ibarra and W. Francke, Orchid pollination by sexual swindle, *Nature*, 1999, **399**(6735), 421.
18. M. Ayasse, J. Stökl and W. Francke, Chemical ecology and pollinator-driven speciation in sexually deceptive orchids, *Phytochemistry*, 2011, **72**(13), 1667.
19. P. R. Ehrlich and P. H. Raven, Butterflies and Plants: A Study in Coevolution, *Evolution*, 1964, **18**, 586.

3 Plants That Eat Insects

3.1 Introduction

The previously-described schemes of plants to attract insects were geared towards pollination and, thus, to the reproduction of plants. This chapter addresses plant mechanisms in which plants still attract insects, but then capture and digest them in order to obtain supplemental nutrients. The break-down products from the insects enable plants to survive in nutrient-poor environments.

A small percentage of flowering plants, less than 0.2% of all angiosperms, supplement their nutrition by trapping and digesting insects. They are the carnivorous or, more specifically, the insectivorous plants. They grow in environments that are low in available nitrogen, phosphorus, and other minerals. Plants in these habitats therefore require additional sources of essential nutrients. Many insectivorous plants grow in water-logged, boggy places with acidic soils, sometimes with a pH of 3–4. Swamps and bogs have a high content of organic materials. But the high acidity suppresses the activity of beneficial bacteria that can break down organic materials into mineral nutrients that plants can absorb. Insectivorous plants are also found on nutrient-poor soils of tropical rain forests where most of the nitrogen and phosphate nutrients are already absorbed by the surrounding biomass. There are insectivorous plants that live suspended in water and others that grow on trees as *epiphytes, i.e.* as non-parasitic plants that grow on other plants. Interestingly, insectivory among plants evolved independently several times among plants, in unrelated plant families. Insectivorous plants are found in

The Chemistry of Plants and Insects: Plants, Bugs, and Molecules
By Margareta Séquin
© Margareta Séquin 2017
Published by the Royal Society of Chemistry, www.rsc.org

nutrient-poor environments on all continents (except for Antarctica), from the tropics to alpine habitats and even in desert areas.[1]

All insectivorous plants are green plants that undergo photosynthesis. They attract insects to their flowers for pollination. But they also capture insects with modified leaves that have evolved into various types of traps. Figure 3.1 shows examples of insectivorous plants with pitchers, such as the tropical pitcher plants (*Nepenthes* sp.), North American trumpet pitcher plants (*Sarracenia* sp.), or the cobra plant (*Darlingtonia californica*). Pitcher-shaped traps attract insects to the upper rims of the pitchers by a combination of structural and chemical lures. Smooth surfaces and downward pointing hairs inside the pitchers make the trapped insects fall into the water reservoir at the bottom of the pitchers. Once the prey is captured, it is digested by enzymes, and the breakdown products are absorbed by the plant as supplemental nutrients. Insect fecal matter, from insects searching for an exit, can be an additional source of nitrogen compounds and minerals. Other insectivorous plants have sticky glands on leaves, as in sundews (*Drosera* sp., Figure 3.2(a)), or slimy leaves that trap insects, as in butterworts (*Pinguicula* sp., Figure 3.2(b)). Both genera are widespread types of carnivorous plants. Some modified leaves can even act as closing traps, as in Venus flytraps (*Dionaea muscipula*), native to the subtropical wetlands of Eastern USA (Figure 3.2(c)).[2]

(a)   (b)   (c)

Figure 3.1 Insectivorous plants with pitchers. (a) Tropical pitcher plant (*Nepenthes* sp.), with anthocyanin pigments. (b) Trumpet pitcher plant (*Sarracenia* sp.), with both pitchers and flowers; designs of anthocyanin pigments lure insects to the pitchers. (c) Cobra plant (*Darlingtonia californica*), with patterns of anthocyanin pigments and semi-transparent pitchers.

(a)   (b)   (c)

Figure 3.2   Insectivorous plants with slimy or sticky modified leaves. (a) A
New Zealand sundew (*Drosera* sp.), with sticky droplets on
leaves. (b) An alpine butterwort (*Pinguicula* sp.), with slimy
leaves and trapped insects on them. (c) Venus flytrap (*Dionaea
muscipula*), with anthocyanin pigments that attract insects to
the open trap.

The theme of insectivorous plants has long captured people's
interest and keeps attracting the public in exhibits and in popular
literature.[3] Insectivorous plants fascinated the naturalist Charles
Darwin enough to write an entire book about them.[4]

3.2   Luring and Digesting Insects for Supplemental Nutrition

Carnivorous plants not only use structural traps and lures; they also
use intricate chemistry to attract insects, trap them, and then finally
digest them for supplemental nutrients.

Chapter 2 addressed the chemistry of compounds that plants
evolved to attract insects to their flowers, like scents, colorful pigments, or offerings of sweet nectar. The same type of compounds lure
insects to the modified leaves of insectivorous plants. Volatile compounds, similar in chemistry to floral volatiles, are often emitted from
the water reservoirs in the pitchers. Sweet, extrafloral nectars lure
insects to the rims of the pitchers of *Nepenthes*, *Sarracenia*, and *Darlingtonia* plants (Figure 3.1). The pitchers also have colorful pigments
and patterns that further attract insects. Similarly, the sticky, glan-
dular hairs of sundews (*Drosera*) and the inner part of the trap in
Venus flytraps (*Dionaea muscipula*) feature showy pigmentations
(Figure 3.2). The purple colors are provided by anthocyanin pigments
(Chapter 2.9). UV-reflective pigments additionally attract insects

to the pitchers. The smooth inner surfaces make insects slip and fall into the water reservoirs at the bottom of the pitchers. Insects drown, and digestion by a complex system of enzymes in the water sets in.

Nepenthes plants can be grown relatively easily in controlled set-ups; thus the chemistry of their pitchers and fluids is well studied. The pitcher fluid is mostly water and contains specific enzymes, *i.e.* biological catalysts that break down insect materials and then promote the absorption of the digestion products by the plants. Recall that enzymes are proteins, *i.e.* polymers of amino acid units, always with defined three-dimensional shapes. Glands on insectivorous plant surfaces not only secrete digestion fluids full of enzymes but can also absorb the released nutrients. Names of enzymes end in –ase and describe the functions that an enzyme performs. A protease breaks down proteins, usually into amino acids, while a lipase breaks down lipids. In pitcher plants and other insectivorous plants there are also chitinases, enzymes that specifically break down the chitin skeleton of insects. Chitin is a polymer of a nitrogenous derivative of glucose. (The structure of chitin will be discussed in detail in Chapter 5.1.) Digestion products from insects spending time in the traps deliver additional supplementary nitrogen, phosphorus, and minerals.[5]

Other insectivorous plants have sticky or slimy layers on their leaves that serve as insect traps (Figure 3.2). The sticky hairs of sundew plants (*Drosera*) covered with glands, the slimy leaves of the alpine butterworts (*Pinguicula*), and the inner surfaces of Venus flytraps (*Dionaea muscipula*) secrete *mucilage* that get insects stuck. Mucilage is a thick, gluey, gel-like substance produced in many unrelated plants (*e.g.* also in the inner part of cactus pads). It consists of a complex mixture of several polymeric substances like polymeric carbohydrates, and also proteins that have carbohydrate chains bonded to them, the so-called glycoproteins. Mucilage secreted from sundews has been identified as a *hydrogel*, consisting of hydrophilic polymer chains that are crosslinked. Mucilage is highly polar due to its many OH groups and can hold large amounts of water. This gives it a slimy quality that traps and drowns insects. Enzymes in the mucilage then start digesting and breaking down the captured insects into nutrients useful to the insectivorous plants.

As an aside, sundew mucilage is a topic of great interest in recent biomedical research due to its adhesive, biocompatible, and biodegradable properties. As a biological adhesive, it has potential applications in tissue engineering, *i.e.* in the repair and reconstruction of mammalian tissues.[6]

(a)   (b)

Figure 3.3 Noncarnivorous plants that hold insects captive in their flowers, then release them. (a) Lady's slipper orchid (*Cypripedium* sp.). (b) Giant pipevine (*Aristolochia gigantea*).

As another facet of the intricate interactions between insects and carnivorous plants, some insects have learned to go around the trapping mechanisms. Some insects have footpads coated with lipids that act as nonstick surfaces that allow the insects to move around on the sticky plant surfaces and feed on the trapped insects themselves.[7] Some spiders take advantage of the trapped insects – and set up their webs right below the pitchers' openings, thus obtaining their own proteins and depriving the plant from obtaining the extra nitrogen.

On the other hand, there are plants that have flowers shaped like traps that capture insects for a while and then release them again, but do not prey on them. While the insects are buzzing around within the flowers, they collect pollen which they then carry on to the next plant. Examples of such flowers are lady's slipper orchids (*Cypripedium* sp.) or the flowers of the tropical pipevine *Aristolochia gigantea* indigenous to Brazilian and Costa Rican rain forests (Figure 3.3). The deep maroon colors of the pipevine, created by high concentrations of anthocyanins, attract dung-loving flies. The insects are held captive in the pouch-shaped flowers, get covered by pollen until released, and then carry it to the next flower.[8]

3.3   Conclusions

Plants have evolved many different mechanisms that allow them to survive in environments with different challenges for plant growth. Insectivorous plants have developed ways to survive in nutrient-poor environments by trapping and digesting insects, somehow reversing the common roles of predator (usually the insects) and prey. Complex interrelationships between insects and carnivorous plants keep evolving, with insects learning to cope with trapping mechanisms, or insectivorous plants and insects forming symbiotic relationships.

Read on about very different survival mechanisms of plants in the next chapter.

References

1. B. A. Rice, Reversing the Roles of Predator and Prey, in *All Flesh is Grass*, ed. J. Seckbach and Z. Dubinsky, *Cellular Origin, Life in Extreme Habitats and Astrobiology*, Springer, Dordrecht, 2011, vol. 16, pp. 493–518.
2. P. Jolivet, *Interrelationship Between Insects and Plants*, CRC Press, Boca Raton, 1998.
3. P. D'Amato, *The Savage Garden*, Ten-Speed Press, Berkeley, CA, 2013.
4. C. Darwin, *Insectivorous Plants*, John Murray, London, 1875.
5. A. Mithöfer, Carnivorous pitcher plants: Insights in an old topic, *Phytochemistry*, 2011, **72**, 1678.
6. Y. Huang, Y. Wang, L. Sun, R. Agrawal and M. Zhang, Sundew adhesive: a naturally occurring hydrogel, *J. R. Soc. Interface*, 2015, **12**, 226.
7. D. Voigt and S. Gorb, An insect trap as habitat, *J. Exp. Biol.*, 2008, **211**, 2647.
8. I. Urru, M. C. Stensmyr and B. S. Hansson, Pollination by brood-site deception, *Phytochemistry*, 2011, **72**, 1655.

4 Plants' Defense Against Insects

4.1 Introduction

Plants cannot move away when insects infest them. Yet, the majority of plants are thriving, in spite of their immobility and masses of hungry insects. This is largely due to numerous chemical compounds of great diversity and often high complexity that plants have evolved in response to insect attacks. Specific organic compounds in leaves, roots, or tree barks, with sticky, distasteful, or toxic properties, make plant parts undesirable as food for insects (and often for other animal browsers as well) and support plants in their struggle for survival. Plants synthesize their chemical defenses from primary metabolites like carbohydrates, amino acids, and fatty acids. These defensive compounds are secondary metabolites and as such not essential for plant growth. But they are nevertheless important for the survival of plants as they discourage insects from feeding on leaves, stems, or roots. The various defensive compounds occur selectively in certain plant families, genera, or species. Most of the plants' repellents do not kill insects. But they render plant parts unpalatable so that insects avoid them or eat only small portions. Therefore, plants can continue to grow.[1]

Different amounts of the chemical defenses are produced during different stages of a plant's life. Young, tender shoots of green plants are often more prone to insect attacks (as every gardener knows) (Figure 4.1(a)). Their leaves tend to contain low concentrations of the chemical defenses, the young plants expending most of their energy producing the basic plant structures. This is often attested by the mild taste of young vegetables, like new leaves of lettuce or dandelion,

The Chemistry of Plants and Insects: Plants, Bugs, and Molecules
By Margareta Séquin
© Margareta Séquin 2017
Published by the Royal Society of Chemistry, www.rsc.org

Figure 4.1   Leaves of angiosperms and insect damage. (a) Young alder leaves (*Alnus* sp.) skeletonized by a leaf beetle. (b) Fossilized angiosperm leaf from a Cretaceous forest, with insect feeding damage. (Photo by Dori Lynne Contreras.)

quite different from the grown-up, fully developed plants that have accumulated defensive substances which provide bitterness. Plants under stress, as in a drought or after a freeze, are weakened and less able to synthesize the deterring compounds.

Fossil finds of ancient plants with traces of insect damage have shed light on the evolution of defensive plant compounds. Plant fossils (Figure 4.1(b)) show a response to the feeding behavior of ancient insects. The patterns and shapes of the damage in different fossilized plant parts, like specific arrangements of holes in leaves or leaf mining, have been compared with those of contemporary herbivorous insects. These studies allow determination of the types of insects that fed on the ancient plants. Up to about 200 million years ago, until the end of the Cretaceous era, fossils of insect-damaged plants show that most insects were *polyphagous*, *i.e.* fed on many different plants. Later fossils, like finds in amber, have indicated a shift towards insects specializing in selected plants as foods. This is seen as a sign that plants started to evolve specific defenses in response to insect attacks. It also suggests the selective deterrent effects of defensive plant compounds.[2] The ongoing coevolution of plants and insects has plants further adapt to insect attacks, just as insects learn to cope with plant defenses.

Several of the chemical families that were shown earlier as attractants show up in this chapter again, but here in their roles as plant defenses. Volatile compounds emitted by leaves can act as insect repellents. In addition, a large and diverse collection of bitter, sticky, or toxic compounds aid plants to survive insect attacks. The following descriptions address some of the major groups of chemical plant defenses.

4.2 Plant Secretions That Trap Insects

Many different, unrelated plants produce sticky, elastic substances as a means of defense. These viscous secretions cover vulnerable plant parts and incapacitate attacking insects. In everyday usage, the sticky substances are commonly – and sometimes interchangeably – known as resins or gums. The following accounts describe their differences, occurrences, and chemical properties.

Our focus is first on resinous exudates that are particularly produced by conifers (order Coniferales). Examples of conifer trees are pines (Figure 4.2(a)), spruces, and firs (Pinaceae), or New Zealand kauri trees (*Agathis australis*, Araucariaceae) (Figure 4.2(b)). Conifers inhabit a large part of the earth's land mass and live in many different habitats. They include the oldest and largest living organisms. Their success is derived largely from a highly developed chemical defense system. This includes viscous secretions that allow conifers to withstand many different challenges to their survival, like attacks by insects. Resinous substances, exuded when part of a tree is damaged, are mobilized by the plant, seal the wound, and get insects stuck.[3]

Trees that are stressed by environmental factors are less capable of synthesizing defensive compounds. As an example, during the ongoing exceptionally long drought period in the western United States (at the time of writing), trees cannot produce enough defensive resins to incapacitate attacking insects. Vast numbers of pines have

(a)　　　　　　　　(b)　　　　　　　(c)

Figure 4.2　Plant resins and insect attacks. (a) Resins ooze out from a cut pine branch. (b) Resins exude from a wounded kauri tree (*Agathis australis*). (c) Drought-stressed pines produce less resins and often succumb to bark beetle attacks.

succumbed to attacks by bark beetles (Figure 4.2(c)). Thus, the trees are being decimated – and provide food for the beetles.[4]

A characteristic property of true resins is their insolubility in water. Resins produced by conifers are composed of oleoresin, commonly known as pitch.[5] It is the viscous, often fragrant secretion that conifers exude in response to wounding. Figure 4.3 shows the chemical structures of some common constituents of resins. Oleoresin is a complex mixture of isoprenoids (compare Chapter 2.6). It contains volatile monoterpenes and sesquiterpenes, like α-pinene (C_{10}) **4.1**, β-pinene **4.2**, and caryophyllene (C_{15}) **4.3**, and less volatile diterpenes like abietic acid (C_{20}) **4.4**. Turpentine (or oil of turpentine) is the liquid fraction of resins and acts as a solvent for higher terpenoids, carrying them to the site of injury. Being mostly composed of hydrocarbons, resins are insoluble in water – and highly flammable, as attested in a forest fire. The volatile mono- and sesquiterpenes

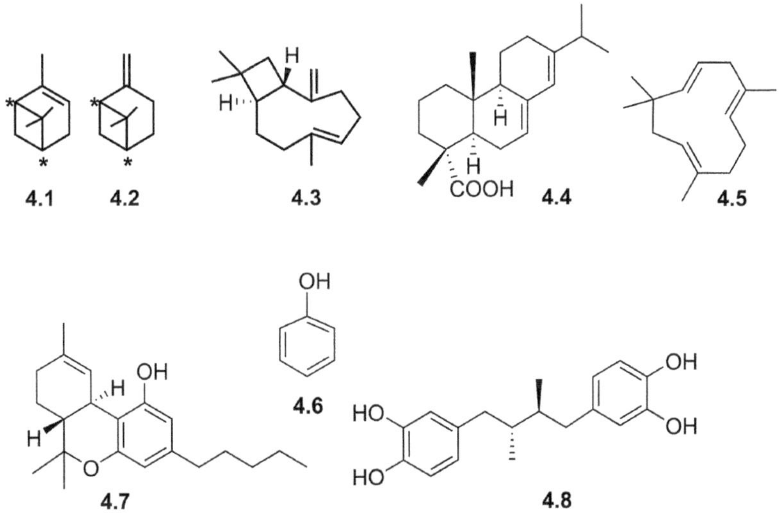

Figure 4.3 Common constituents of resins: terpenes and phenolics. The chiral carbon atoms in the terpene structures of α-pinene (C_{10}) **4.1** and β-pinene **4.2** are pointed out by asterisks (*). The three-dimensional arrangements around chiral centers in the terpene structures of caryophyllene (C_{15}) **4.3** and abietic acid (C_{20}) **4.4**, as well as in the phenolic structures of tetrahydrocannabinol (THC) **4.7** and nordihydroguaiaretic acid (NDGA) **4.8**, are defined by using bold bonds (pointing towards the reader) and broken bonds (oriented away from the reader). Humulene **4.5** is a sesquiterpene without chiral carbons. The phenol structure **4.6** is part of phenolic molecules.

contribute to the typical "pine" smells of resins, especially on a warm day. When exposed to air, the volatiles evaporate, while the less volatile diterpenes polymerize due to oxidation in air and harden, forming a protective layer over the injured plant parts. At higher ambient temperatures resins have a viscous liquid consistency that traps insects. Resins are produced in special cells that secrete the resins into resin ducts, usually in response to injury. Fossilized resins from conifers were the source of amber, and specimens of amber have provided information on ancient insects and plants.

Besides terpenoids, resins can also contain phenolics (Figure 4.3), with a phenol structure **4.6** being part of their molecules. They are especially found in plants other than conifers. Deciduous trees and bushes, like willows (*Salix* sp.), cottonwoods (*Populus* sp.), or alders (*Alnus* sp.), protect their young twigs and buds with viscous resins. Herbaceous plants like hop (*Humulus lupulus*) (Figure 4.4(a)) or the marijuana plant (*Cannabis sativa*) (Figure 4.4(b)) are well-known for their resinous leaf coverings. The sesquiterpene humulene $C_{15}H_{24}$ **4.5** from hop and the phenolic tetrahydrocannabinol THC **4.7** of cannabis are components of the resins. Creosote bush (*Larrea tridentata*), a widespread desert plant from Mexico and Southwestern USA, got its name from its sharply-scented phenolic resins that protect leaves against insects as well as against water loss. The resins contain the phenolic nordihydroguaiaretic acid (NDGA) **4.8**.

(a)   (b)   (c)

Figure 4.4 Examples of herbaceous plants that produce glabrous protective layers. Leaves of (a) hop (*Humulus lupulus*) and (b) cannabis (*Cannabis sativa*), both in the Cannabaceae family, are covered with resins. (c) The flower buds of gumweed (*Grindelia* sp.) are protected by sticky gums.

Gums are different, sticky, elastic plant products, although in everyday usage they are often called resins, and *vice versa* resins are commonly addressed as gums. By definition, resins do not dissolve in water and are composed of terpenoids or phenolics or mixtures of them, while true gums are hydrophilic polysaccharides. The gum-diggers of New Zealand that collected "gum" from kauri trees actually harvested resin produced by the trees. Blue "gum" eucalyptus trees (*Eucalyptus globulus*), native to Australia and widely planted in many parts of the world, produce fragrant resins containing terpenes and phenolics. True gums consisting of polysaccharides form protective layers in many herbaceous plants. They also form sticky, hydrophilic secretions when trees of the genus *Prunus* (like cherry, almond, or plum trees) are wounded. For example, viscous, semiliquid gums cover surfaces of freshly cut branches of cherry trees. Insect infestations, adverse weather conditions, or mechanical damage cause gummosis, a defense mechanism of fruit trees of the genus *Prunus* that leads to the deposit of patches of gum on the bark of trees. Protective gums can also cover vulnerable parts of smaller herbaceous plants, like the flower buds of gumweed (*Grindelia* sp.) shown in Figure 4.4(c). Later in this chapter we'll encounter latex, *e.g.* as a source of rubber, which is sometimes also confused with gums. Sticky plant substances were also featured in the chapter on insectivorous plants in the form of polysaccharide-based mucilage. To complicate matters further, some plants produce both gums and resins, or mixtures of them. But a common characteristic of the diverse, sticky, elastic plant products is their protective role against insect attacks and in challenging environmental situations like drought.

4.3 Deterring Volatiles from Leaves

Volatile organic compounds from plants have many different roles in plant–insect interactions. In Chapter 2 we encountered examples of terpenes, alcohols, aldehydes, and esters as floral volatiles that attract pollinating insects. The same compounds can act as insect deterrents in leaves or roots, keeping away herbivores. Some volatiles released by plants can also signal insect infestations to neighboring plants, leading to the production of defensive compounds there, or can attract other insects that act as predators on the invading herbivores.

Leaves with strong scents have long been used to keep away unwanted insects. The old saying, "Leaves of bay keep the weevils away", implies that bay leaves (*Laurus nobilis*, Figure 4.5(a)), with their typical

(a)   (b)   (c)

Figure 4.5   Plants that give off insect-deterring volatiles from their leaves.
(a) Bay (*Laurus nobilis*). (b) Lemongrass (*Cymbopogon citratus*).
(c) Branches of *Eucalyptus* sp.

odor due to their essential oils, are a deterrent to insects like weevils (small insects of the Curculionidae family) that might invade food items like flour and grains. Gardeners often plant strongly scented plants like lavender (*Lavendula* sp.), rosemary (*Rosmarinus officinalis)*, or mints (*Mentha* sp.) around rose bushes to reduce insect infestations of their roses. Citronella oil, obtained from lemongrass (*Cymbopogon* sp., Figure 4.5(b)), is commonly used as an insect-repellent.[6]

Some of the insect-repelling compounds in these plants are the monoterpenes linalool **2.9** from lavender leaves, and citronellol **2.10** and limonene **2.11** found in lemongrass. (They were encountered earlier, in Chapter 2, as floral attractants towards distinct pollinating insects.) A further example of an insect-repelling terpene is 1,8-cineole or eucalyptol, **4.9** (Figure 4.6), with a camphor-like smell. It is a major component of the essential oils of fresh rosemary, bay leaves, and eucalyptus oil from *Eucalyptus* spp. (Figure 4.5(c)). Pulegone **4.10**, occurring in mints (*Mentha* sp.), has a strong scent of spearmint. Note that molecules of pulegone have a chiral center. Figure 4.6 shows the enantiomer that is naturally occurring in mints.

Undamaged leaves release small amounts of volatiles from leaf surfaces. But greater quantities are emitted when herbivores attack the plants. In response to injury plants synthesize defensive volatiles *de novo*. Some of the volatiles are released shortly after damage by feeding insects (or other mechanical damage). Green-leaf volatiles, with the smell of freshly cut grass, are emitted when leaves or grass blades are damaged (as in mowing the lawn) or bitten into by herbivores. The volatiles are biosynthetically formed from fatty acids

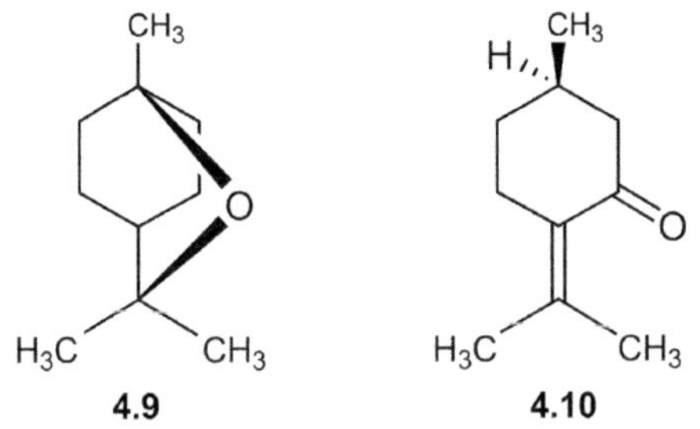

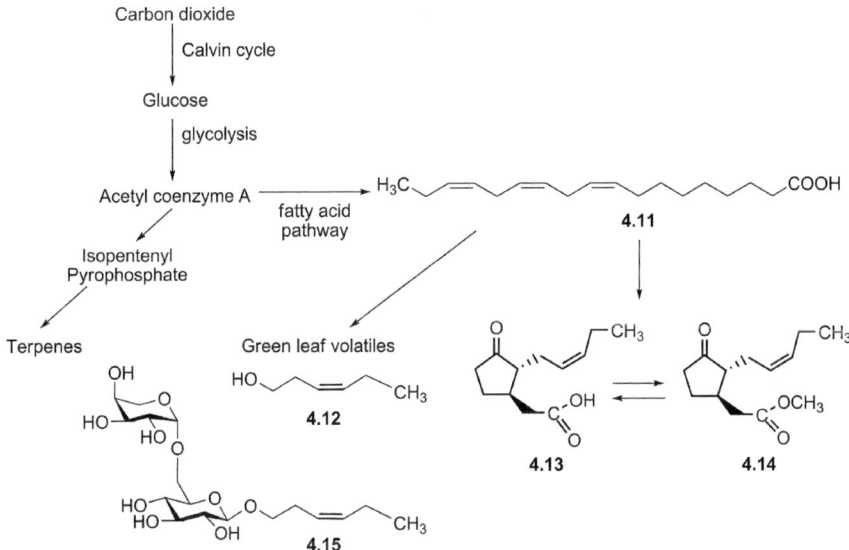

4.9   **4.10**

Figure 4.6 Examples of insect-repelling monoterpenes. 1,8-Cineole (eucalyptol) **4.9** is a common insect-repelling terpene found in rosemary, bay leaves, and *Eucalyptus* spp. Pulegone **4.10** is a chiral compound; the enantiomer found in mints (*Mentha* sp.) is shown.

Figure 4.7 Simple scheme of biosynthetic pathways that lead to defensive plant volatiles. The metabolic conversion of linolenic acid **4.11** to the green-leaf volatile (Z)-3-hexen-1-ol **4.12**, the plant hormone jasmonic acid **4.13**, and its derivative methyl jasmonate **4.14** is shown. The vicianoside **4.15**, a glycoside of **4.12**, has been found to negatively affect the development of herbivorous cutworms.

like linolenic acid **4.11** (Figure 4.7). Green-leaf volatiles are structurally related to hexanol, with six carbon atoms. A very common green-leaf volatile in herbaceous plants is (Z)-3-hexen-1-ol **4.12**. These volatiles can alert insect predators and parasitoids (such as parasitic wasps) who can then distinguish between infested and noninfested plants and locate insect prey, thus actually helping the plants. There are many examples of *multipartite relationships* that link a plant, its herbivores, and the herbivores' natural enemies.[7]

Within about 24 hours of leaf damage, many plants produce additional volatile defenses. In response to tissue damage, the plant hormone jasmonic acid **4.13** (Figure 4.7) and its volatile derivative methyl jasmonate **4.14** can form which in turn trigger the increased production and release of defensive volatile organic compounds, like mono- and sesquiterpenes. Figure 4.7 shows a simple scheme of common biosynthetic pathways that lead to the synthesis of the volatiles. Glucose, formed from the Calvin cycle during photosynthesis, undergoes glycolysis which in turn leads to the key compound acetyl coenzyme A. From there, fatty acids are formed by way of the fatty acid pathway. Linolenic acid is one of the origins of jasmonates and green-leaf volatiles. Different pathways lead from acetyl coenzyme A to the synthesis of terpenes.[8,9]

Furthermore, plant volatiles can act as signals towards uninjured plants and induce defense mechanisms in them. A recent study involved tomato plants infested by cutworms (*Spodoptera litura*). These moth caterpillars can destroy tomato plants within a short time. An attack by cutworms caused leaves of tomato plants to release (*Z*)-3-hexen-1-ol **4.12** which was picked up from the air by uninfested plants and converted by them to the glycoside (*Z*)-3-hexenyl-vicianoside **4.15**. The latter compound impedes the maturation and survival rate of the cutworms. This example illustrates the complexity and interconnectedness of mechanisms that involve defensive plant volatiles released in response to herbivore attacks.[10]

Another example of a multipartite plant–insect relationship involves *Cotesia marginiventris*, a parasitoid wasp that attacks many types of plant-feeding caterpillars. The wasp is attracted to caterpillars feeding on maize varieties (*Zea mays*) by caterpillar-induced plant volatiles. The wasp thus provides an indirect protection for the plants.[11]

Roots produce defensive volatiles, too, and many of them are similar to leaf volatiles. Roots release the volatile defenses when underground insects, like grubs, start feeding on them. In a study of volatiles emitted by barley roots (*Hordeum vulgare*), twenty-nine different volatile organic compounds were identified, among them mostly fatty acid derivatives like (*Z*)-3-hexen-1-ol.[12] Root volatiles are much less studied than defensive volatiles from leaves.

About two thirds of all insects are herbivores, and many of them feed on crop plants important for human food production. Therefore, studies of the mechanisms of the respective plant–insect interactions are of great interest and topics of active research; they frequently support efforts to find effective methods to control insect pests. Mechanisms by which plants react to tissue injury are diverse and intricate, and much still has to be discovered about them. Different types of herbivores result in the emission of different volatiles. Oral secretions from chewing insects often contain enzymes and can influence which plant volatiles are produced. The final outcome of the plant response also depends on the developmental stage of the plant and the surrounding environmental conditions.

Producing the insect deterrents is costly in metabolic terms for the plants and requires considerable energy from plant systems. Stressed plants are less capable of producing the defensive compounds. Meanwhile, herbivorous insects have adapted – and keep adapting – to resist the plant defenses. This requires metabolic costs from the insects as well. As a consequence, both hosts and herbivores survive in most plant–insect interactions, but their developments are not optimal.

4.4 Cyanide in Response to Insect Attacks

Cyanogenic glycosides are defensive compounds that generate hydrogen cyanide when leaves or other plant parts that contain them are damaged. The bitter compounds are stored in the cells of intact plant tissues. When herbivorous insects start feeding, plant tissues are destroyed and enzymes stored in other parts of the cells react with the cyanogenic glycosides to induce the formation of toxic cyanide. Cyanogenics are widespread in plants; they are found in unrelated families in more than 2500 plant species. They occur *e.g.* in bitter almonds, and in the pits of peaches, cherries, and apples, all plants of the rose family (Rosaceae). (If we bite into apple seeds, we break plant cell walls and experience the bitter taste of the cyanoglycosides and their hydrolysis products.) Their wide occurrence indicates that they are ancient plant defenses. Figure 4.8 shows some important crop plants used for human consumption that contain these plant defenses. Examples are cassava roots, also known as tapioca, manioc, or yuca (*Manihot esculenta*, Euphorbiaceae, Figure 4.8(a)), a native of South America, or the ancient African grain sorghum (*Sorghum bicolor*, Figure 4.8(b)) in the grass family (Poaceae), or lima beans

Figure 4.8 Plants from diverse plant families contain cyanogenic glycosides. (a) Cassava, also known as tapioca, manioc, or yuca (*Manihot esculanta*), is a plant in the family of Euphorbiaceae. (b) Sorghum (*Sorghum bicolor*) is in the grass family (Poaceae). (c) Lima beans (*Phaseolus lunatus*), here shown shelled, belong to the pea family (Fabaceae).

(*Phaseolus lunatus*, Figure 4.8(c)) in the pea family (Fabaceae). White clover (*Trifolium repens*, Fabaceae), a widespread pasture plant in many European countries, has been extensively studied for its content of cyanogenic glycosides.[13]

The general structure of cyanogenic glycosides **4.16** is shown in Figure 4.9. R_1 and R_2 signify attached groups that can be identical or different. In linamarin **4.17**, the two groups, R_1 and R_2, are both CH_3 (methyl) groups. Linamarin is found in the seed skins of flax (*Linum usitatissimum*). (Chemical common names often recall the occurrence of the compounds in typical plants.) Linamarin is also the major cyanoglycoside in lima beans and in leaves and tubers of cassava (tapioca). In intact plant tissues, cyanogenic glycosides are stored in the plants' cell vacuoles. Note that the sugar part, shown in detail in the structure of linamarin, has numerous OH groups, making the compounds water-soluble in the aqueous environment of the vacuoles. If R_1 and R_2 groups are different, as in the examples of amygdalin **4.18** and prunasin **4.19**, the molecules have four different groups attached to the central carbon atoms bonded to the cyano groups (CN). This means that amygdalin and prunasin, like most of the cyanogenic glycosides, are asymmetric or chiral. The structures of amygdalin and prunasin both show the 3D arrangement of the groups. Amygdalin and prunasin, two common cyanogenic glycosides, both occur in bitter almonds. The enantiomer of prunasin shown, named sambunigrin, is typically found in elderberries

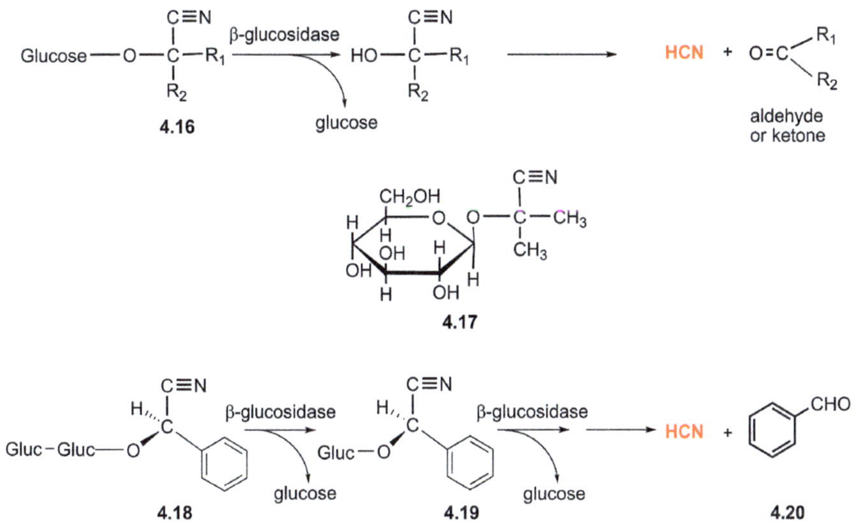

Figure 4.9 Cyanogenic glycosides. General structure of cyanogenic glycosides **4.16**. The reaction with β-glucosidase, followed by hydrolysis upon tissue damage, leads to the formation of HCN and an aldehyde or ketone. Linamarin **4.17**, amygdalin **4.18**, and prunasin **4.19**, are common cyanogenic glycosides, the latter two forming benzaldehyde **4.20** and HCN upon reaction with β-glucosidase.

(*Sambucus* sp.). Cyanogenic glycosides are biochemically derived from amino acids, *e.g.* prunasin and amygdalin from phenylalanine **2.23**.[14]

When plant tissues are damaged by feeding herbivores, the glycosides are exposed to β-glucosidases that are located in other parts of the cells. Interaction with these enzymes induces the breakup of the cyanogenic glycoside. The reactions result in the formation of glucose and an intermediate that spontaneously decomposes into toxic hydrogen cyanide (HCN) and a ketone or an aldehyde (Figure 4.9). As an example, the breakdown of amygdalin and prunasin due to β-glucosidase produces benzaldehyde **4.20**, with an almond smell, and HCN. Hydrogen cyanide is well known for its toxicity, to humans as well as to herbivores.

As an aside, plant foods with cyanogenics that are used for human consumption, like sorghum, cassava, or lima beans, require proper preparation. Milling or cooking of the plant materials destroys plant tissues. In turn, β-glucosidase enzymes destroy the cyanogenic glycosides. HCN, a gas, is released and dissipates, and the plant foods become edible for humans. Meanwhile, the growing crop plants have the desirable insect defense.[15]

Cyanogenic glycosides are not only toxic, but have a bitter taste that repels insects keen on feeding on the plants. Nevertheless, some insects have evolved mechanisms to get around this toxicity, feeding on cyanoglycosides, sequestering them, and even taking advantage of nutritious glucose produced. It may be surprising that some insects are capable of synthesizing cyanoglycosides themselves *de novo*. Examples will be addressed in Part 2, the Insect Perspective, in Chapter 8.3.

4.5   Glucosinolates and Pungent Volatiles

Glucosinolates, also known as mustard oils, are sulfur- and nitrogen-containing plant compounds. They are found mainly in plants of the cabbage family (Brassicaceae) as well as the caper family (Capparidaceae). These defensive compounds are stored in cells throughout the plants. Comparable to the mechanisms of cyanogenic glycosides, damage to plant tissues, *e.g.* from feeding insects, leads to the reaction of enzymes with the glucosinolates. These reactions form pungent, insect-repelling volatiles. Glucosinolates are mostly re-stricted to the plant families named above. At this point more than 120 different structures of mustard oils are known. Figure 4.10(a) and (c) show examples of plants that contain them, like all types of cabbages (*Brassica oleracea*) or the caper bush (*Capparis spinosa*). Many agriculturally important crop plants contain glucosinolates. Thus, their interactions with herbivorous insects, like the cabbage white butterfly (*Pieris brassicae*, Figure 4.10(b)), have been studied extensively.[16]

(a)							(b)							(c)

Figure 4.10   Plants that contain glucosinolates and insects adapted to them. (a) All types of cabbages (*Brassica oleracea*) contain glucosinolates. (b) Cabbage white butterfly (*Pieris brassicae*). (c) Caper bush (*Capparis spinosa*) with buds and flower.

Figure 4.11 shows the general structure of glucosinolates 4.21. They contain a thioglucose part ('thio' signifying that a sulfur atom replaces the oxygen atom in the glucose linkage), an ionic group, and a variable side chain R. Glucosinolates are stored in the plants' cell vacuoles as their glucoside components, together with the ionic groups, make them hydrophilic compounds. The enzyme myrosinase is stored in specialized cells in all plant parts. When plant tissues are disrupted, myrosinase comes into contact with the glucosinolates that, as a consequence, are hydrolyzed. The products of these interactions are glucose, sulfate, and characteristic volatiles with a pungent odor, like isothiocyanates (R–N=C=S) 4.22, that are also toxic to many insects. Figure 4.11 shows the general reaction of the hydrolysis of glucosinolates leading to the repelling volatiles. Sinigrin 4.23 is a common glucosinolate in the Brassicaceae found, *e.g.* in horseradish or in seeds of black mustard. Its hydrolysis produces allyl isothiocyanate 4.24, a volatile compound with a sharp smell and taste that people are quite familiar with. Glucosinolates are biosynthesized from amino acids, with methionine 2.34 being a common precursor.[17]

There are numerous examples of insects that have learned to tolerate and to adapt to glucosinolates, in classic cases of coevolution between plant defenses and insects. Many adapted herbivores have become serious pests on crop plants. The caterpillars of the cabbage white butterfly (*Pieris brassicae*, Figure 4.10(b)), the larvae of the cabbage fly (*Delia radicum*), and aphids like green peach aphids (*Myzus persicae*) all feed voraciously on plants of the cabbage family. Different mechanisms enable the insects to tolerate the defensive plant compounds. In *Pieris brassicae*, the hydrolysis of the glucosinolates is redirected towards the formation of less toxic nitriles (R–C≡N) which the larvae can excrete. Female cabbage white butterflies not only tolerate the glucosinolate sinigrin, but use it as a stimulant for oviposition (laying eggs). Then again, different insects may prey on the larvae, undeterred by emitted isothiocyanates. Glucosinolates often induce *multitrophic interactions, i.e.* interactions that involve several levels of organisms in the food chain, including the host plant that produces the defensive compounds, several herbivorous insects adapted to them, and natural parasites that prey on the herbivores.

Concentrations of glucosinolates can vary greatly within different plant parts, often with high concentrations in the seeds, as in mustard seeds. They can affect how desirable crop plants are for human consumption. Thus, studies of plants that contain them and how they resist insect pests, as well as respective insect adaptations and insect predators, are topics of active research. Plant breeders are most interested in learning how concentrations of glucosinolates in crops affect insect resistance.

Glucosinolates are examples of *semiochemicals, i.e.* natural chemical compounds that act as communication between living organisms and affect their behavior, like oviposition in the cabbage white butterfly. Communications can be inter-specific as between a plant and an insect (or two different types of insects), like the examples above. (We shall later encounter examples of intra-specific semiochemicals, in the form of insect pheromones, in Chapter 5.2.)

4.6   Diversity of Bitter-tasting Insect Repellents

Plants have evolved a multitude of defensive compounds that to humans have a distinct bitter taste and, as we can observe, deter insect herbivores. The bitterness of humulene **4.5** in hop plants and of cyanogenic glycosides was mentioned earlier. Bitter-tasting compounds are widespread in plants. Because of the great structural diversity of the compounds, they serve here as a review of previously encountered chemical families of natural products as well as an introduction to some additional categories. While we cannot know the exact experience that insects have, we can observe how they react to the compounds. Insects are generally repelled by bitter compounds.

(a)   (b)   (c)

Figure 4.12   Plants with bitter compounds. (a) Branches of neem plant (*Azadirachta indica*). (Photo by Sushila Kanodia.) (b) Willow (*Salix* sp.). (c) Bark harvested from *Cinchona officinalis*. (Photo by H. Zell. Wikimedia Commons. https://upload.wikimedia. org/wikipedia/commons/0/08/Cinchona_officinalis_001.JPG (accessed: October 2016).)

Figure 4.12 shows examples of insect-repelling plants that contain bitter compounds.

Since ancient times people have observed that insects do not feed on certain plants. (The perceptive gardener notices this, too.) A famous example is the neem tree (*Azadirachta indica*, Figure 4.12(a)), a large evergreen tree native to India and the Indian subcontinent and grown in tropical and semitropical areas all over the world. Its leaves and oil from its fruits and seeds have been used for their insect-repelling (as well as medicinal) properties for more than 2000 years and continue to be used as a natural pesticide today. Neem has a bitter taste. Its bitterness is mainly due to azadirachtin **4.25** (Figure 4.13), a compound with a complex structure indeed. The story of neem serves here as an example of the many steps that led to the definition of its bitter principle. It began with the observation of its insect-deterring properties, which was followed by the isolation of the main bitter compound, then the elucidation of the structure of azadirachtin, and finally the challenging synthesis of the compound. It is a characteristic tale of the numerous steps that are involved in determining the structure of a natural product.

During an invasion of locusts in the Sudan in 1959, the entomologist Heinrich Schmutterer observed that the neem tree was the only plant that remained relatively undamaged while all other vegetation was consumed.[18] The pure compound of azadirachtin, responsible for most of the bitter taste of neem, was isolated from seeds of the neem tree in 1968. The elucidation of its structure was

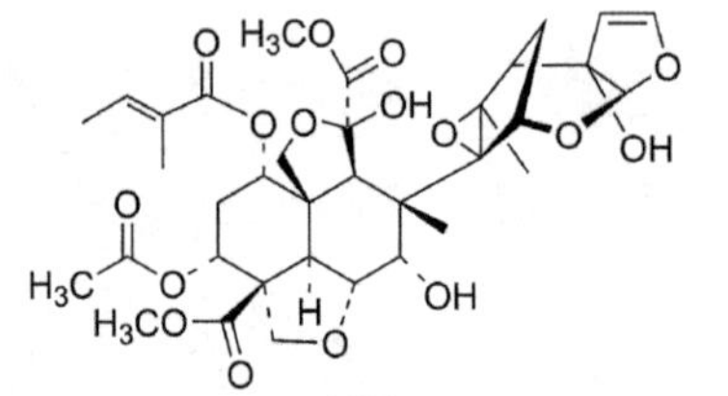

Figure 4.13 Diversity of bitter plant compounds. Azadirachtin **4.25**, a complex terpenoid from the neem tree (*Azadirachta indica*), has strong insect-repelling properties. Examples of insect-repelling phenolics are salicin **4.26**, from willow branches (*Salix* spp.), and the large family of tannins, of which gallic acid **4.27** and the structure of a complex tannin **4.28** are shown. Solanine **4.29** and quinine **4.30** are examples of bitter, insect-repelling alkaloids. The steroid structure of solanine is shown in bold.

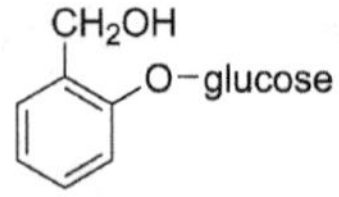

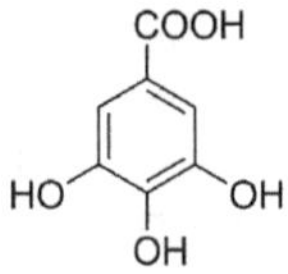

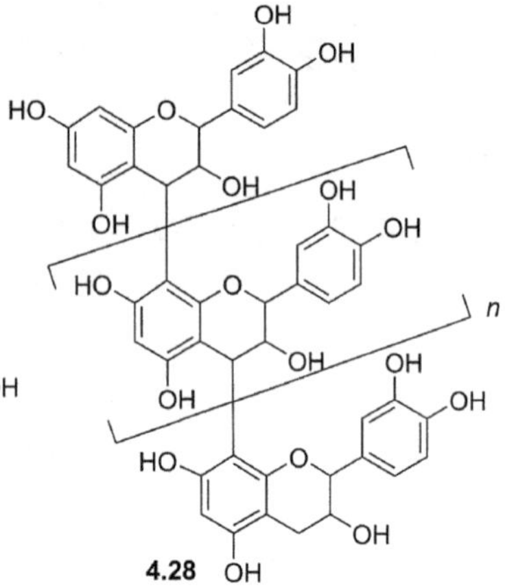

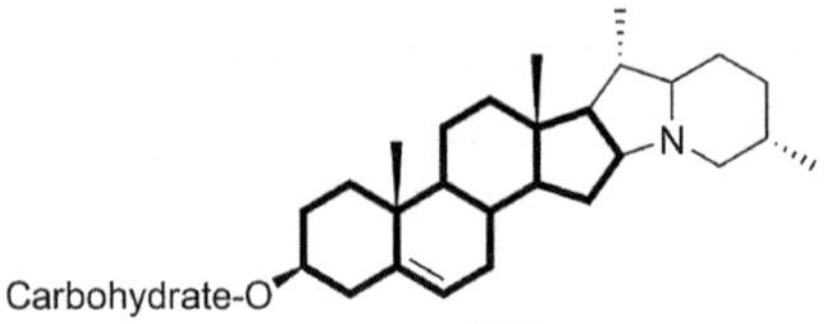

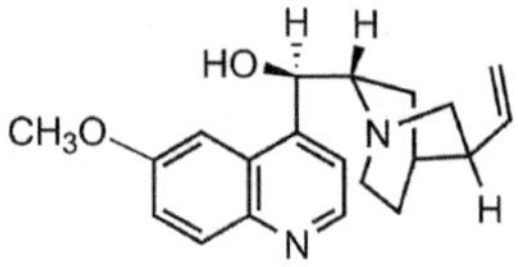

completed in 1987, a challenge that involved many renowned chemists. Azadirachtin, $C_{35}H_{44}O_{16}$, is derived from a terpenoid structure. It contains multiple chiral centers (16!). The ultimate challenge was to synthesize the compound from simple organic compounds, in what is known as the *total synthesis* of a natural product. It took 47 steps and 22 years to accomplish and provided many new insights into chemical reactions that might lead to the synthesis of analogous smaller molecules with similar anti-insect properties. Azadirachtin not only deters insects, but acts also as a growth disruptor against more than two hundred insect species. Most importantly, azadirachtin, while a powerful insect-repellent, has low toxicity towards mammals.[19] A simpler terpenoid with a bitter taste is humulene **4.5** from hop.

Figure 4.13 shows additional examples of bitter-tasting compounds, from different chemical families. Aside from terpenoids there are many phenolics that are insect-deterrents. Willow branches (Figure 4.12(b)) contain the phenolic compound salicin **4.26** that repels herbivorous insects. Yet, there are several species of aphids that infest willows, having adapted to the chemical defenses. These insects are capable of hydrolyzing salicin and make use of the glucose obtained.[20] Tannins are a large family of acidic, bitter-tasting, astringent phenolics. Wood with a high tannin content is insect-repelling – and is prized as lumber because of this quality. Tannins can have simple structures, like gallic acid **4.27**, but most commonly feature complex phenolic structures, with flavonoid components being part of them. An example of a complex tannin is shown in structure **4.28**. The multiple OH groups make tannins hydrophilic. Aside from their bitter taste, tannins form strong bonds with proteins and precipitate them, making them clot or 'denatured'. Enzymes in the gut of insects are peptides and proteins. Thus, they are denatured by tannins and cannot function any more, to the detriment of feeding insects. (A human activity and practical application of tannins denaturing proteins involves "tanning of leather" where gelatinous fresh animal hides are treated with solutions of high tannin content, resulting in hardening of the hides.)

Our selection of structures of bitter-tasting, insect-repelling compounds concludes with examples from the large family of alkaloids which we'll examine in more detail for their toxicity later in this chapter. The alkaloid solanidine **1.2** and its glycoside solanine **4.29** are bitter substances in greening potatoes (yet do not repel the adapted potato beetle). The nitrogen in their structures makes them alkaloids. In addition, they are steroids, with the characteristic steroid structure highlighted in solanine **4.29**. An extremely bitter natural substance, with strong insect-deterring properties, is the alkaloid quinine **4.30** from the bark of the cinchona tree (*Cinchona* sp., Figure 4.12(c)), probably best known for its antimalarial properties.

Some of the compounds above may be better known to us for their human uses. Yet plants mostly evolved them for their defense, namely for keeping herbivorous insects at bay.

4.7 Plant Toxins That Kill Insects

Aside from mere deterrence, many plant compounds harm and kill insects that venture to feed on plant parts. Whether a compound kills or not is a matter of the chemistry of the toxic compounds, their interaction with vital insect organs, and their dose, *i.e.* the amount ingested per feeding. The toxic effect is different for every insect, for every living organism in general. Through evolution, many insects have developed mechanisms that allow them to tolerate the plant toxins and feed on the plants unharmed, as some examples below will illustrate.

Nicotine is a potent natural insecticide. It occurs in all plants of the genus *Nicotiana* (tobacco plants) and some additional plants in the nightshade family (Solanaceae). Figure 4.14(a) and (b) show two different views of the chemical structure of nicotine **4.31**. We use here the example of nicotine for a more detailed look at alkaloids. They are a large family of bioactive natural products, with more than 16 000 known compounds to date.[21,22] Alkaloids are secondary metabolites that mostly occur in plants. They have no known metabolic functions and have probably evolved in defense against insect herbivores. They are well-represented in some plant families, like the nightshade family (Solanaceae), the pea family (Fabaceae), and the poppy family (Papaveraceae), but occur also in other plant families. Alkaloids always contain nitrogen in their structures, with amino acids being their biochemical precursors (Figure 4.15). Their chemical structures characteristically contain rings. Thus, alkaloids are commonly categorized according to their ring structures, sometimes also based on their amino acid origins. Most alkaloids are weakly basic or alkaline. They are known for their bitter tastes

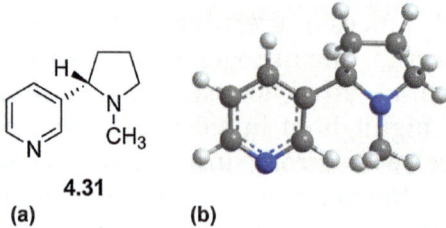

4.31

(a)   (b)

Figure 4.14   Nicotine, a potent insecticide. (a) Molecular structure of the alkaloid nicotine **4.31**. (b) Three-dimensional representation of a nicotine molecule, with the nitrogen atoms shown in blue.

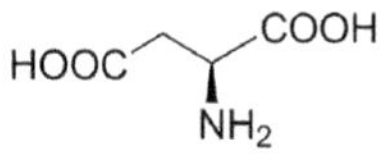

4.32 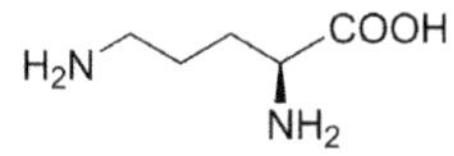   **4.33**

Figure 4.15 Biochemical amino acid precursors of nicotine.

and for their insect-deterring properties. Most alkaloids repel many different insect species at concentrations over 0.1% w/w. Previously we encountered the alkaloid examples of solanine **4.29** and quinine **4.30**.

The structure of nicotine (Figure 4.14(a) and (b)) is composed of two nitrogen-containing rings: an aromatic ring and a five-membered non-aromatic ring, called a pyrrolidine ring. Thus, in alkaloid nomenclature, nicotine is categorized as a pyrrolidine alkaloid. In its 3D-representation (Figure 4.14(b)), the even distribution of the electrons in the aromatic ring is represented by a broken line. Cyclic structures that have other types of atoms besides carbon forming the ring are generally known as *heterocycles*. Note the chiral center in the nicotine molecule. (Try to locate it also in the 3D-structure.) The figures show the naturally occurring enantiomer of nicotine.

Some insects are not deterred by nicotine in plants. The tobacco hornworm caterpillar (*Manduca sexta*) is a major insect pest on tobacco plants. So are several species of aphids that infest the plants.[23]

The amino acid precursors of nicotine are L-aspartic acid **4.32** and L-ornithine **4.33**. Their structures are shown in Figure 4.15. The descriptor 'L' defines the orientation of the chiral centers.

Figure 4.16 shows lupine plants (*Lupinus* spp.) and the structure of lupinine **4.34**, one of the toxic alkaloids typically found in lupines. These bitter alkaloids occur in all lupine plant parts, but are highly concentrated in the seeds. Lupine seeds are not only toxic to insects, but to larger animals as well because of the alkaloids. If used as animal feed the seeds therefore have to be screened for their alkaloid concentration. Lupinine molecules contain a heterocyclic structure called a quinolizidine; two fused six-membered rings, with a nitrogen atom being part of the two rings. Quinolizidine alkaloids are common in lupine plants as well as in plants of the genus *Genista* (French broom or Scotch broom). As Figure 4.16(c) shows, the caterpillars of the genista broom moth (*Uresiphita reversalis*) are not deterred by the alkaloids and may even obtain their own chemical protection from them, in another example of adaptation to plant defenses.

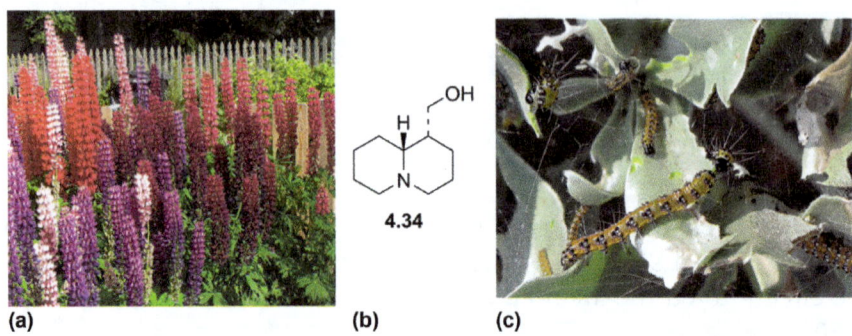

(a)   (b)   (c)

Figure 4.16 Lupine alkaloids and insects that have adapted to them. (a) Lupine plants (*Lupinus* sp.). (b) Lupinine **4.34** is a toxic quinolizidine alkaloid commonly found in *Lupinus* spp. (c) Caterpillars of genista broom moths (*Uresiphita reversalis*) feeding undeterred on a *Lupinus* sp.

4.8   Plant Latex and Its Many Defensive Components

Plant latex, a liquid of milky appearance, occurs in about 10% of all angiosperms, in many unrelated plants. It is a characteristic of plants in the spurge family (Euphorbiaceae), of which the rubber tree (*Hevea brasiliensis*) may be the best-known example. Also plants of the milkweed family (Asclepiadaceae), dogbane family (Apocynaceae), and others, typically contain latex, the common names often alluding to the milky exudates. The word lettuce originates from the French 'laitue', derived from the French word 'lait' for milk, because of the latex in lettuce plants (*Lactuca sativa*, Asteraceae) (Figure 4.17(a)). Latex is mostly a white, non-transparent fluid, but can also have a yellow or orange color. Plants store it under considerable pressure in specialized canals, the 'laticifers'. When plant tissue is damaged by insects, latex is immediately transported to the site of damage. Exposure to air causes it to coagulate, sealing the plant injury, and also gluing up the herbivores' feeding apparatus. Both the coagulating properties of latex and also the toxic compounds in it affect insects adversely. Latex is a complex emulsion of rubber particles (a polymer discussed in detail below), water-insoluble oils, bitter substances, toxins, and enzymes in an aqueous medium. Concentrations of defensive compounds in latex can be more than fifty times higher in the laticifers than in the leaf tissues of a plant. Puncturing a plant vein releases the pressure of latex, leaving other parts of the leaf with

(a)

(b)

Figure 4.17   Latex in plants. (a) Cut lettuce plant (*Lactuca sativa*) showing latex. (b) A drop of white latex exudes from the punctured midvein of a leaf of California milkweed (*Asclepias californica*) infested by oleander aphids (*Aphis nerii*).

much lower concentrations of the chemical defenses. Adapted insects have learned to puncture a main plant vein and then feed on other parts of the leaf. Figure 4.17(b) shows the telltale white drop of latex from a punctured midvein of an aphid-infested milkweed leaf (*Asclepias* sp.).[24] Latex is always present in leaves and stems of the respective plants, but all plant parts can contain it, including the roots.[25]

Plant latex contains defensive compounds from diverse chemical families most of which appeared in previous discussions, like terpenoids (*i.e.* structures derived from isoprene units), alkaloids, and phenolics. In addition, cardiac glycosides are typically part of the latex composition of plants in the dogbane family (Apocynaceae) and the milkweed family (Asclepiadaceae). Various types of enzymes also occur in the milky mixtures. Figure 4.18 shows examples of several components of latex.

A common component of latex is rubber, with tiny particles of rubber suspended in an aqueous medium. They produce the stickiness and white color of latex. The concentration of rubber can be high, as much as 44% in fresh latex in the rubber plant (*Hevea brasiliensis*). Rubber **4.35** is a polymer with a terpenoid structure. Note the repeating isoprene units in the rubber molecule in Figure 4.18. Observe also the specific connection of the monomers. Natural rubber is chemically known as *cis*-1,4-polyisoprene. It is found in 300 plant genera and 8 different plant families. The primary function of latex is one of defense against insects. Mechanisms of the stickiness of rubber particles are not well understood. It has been suggested that

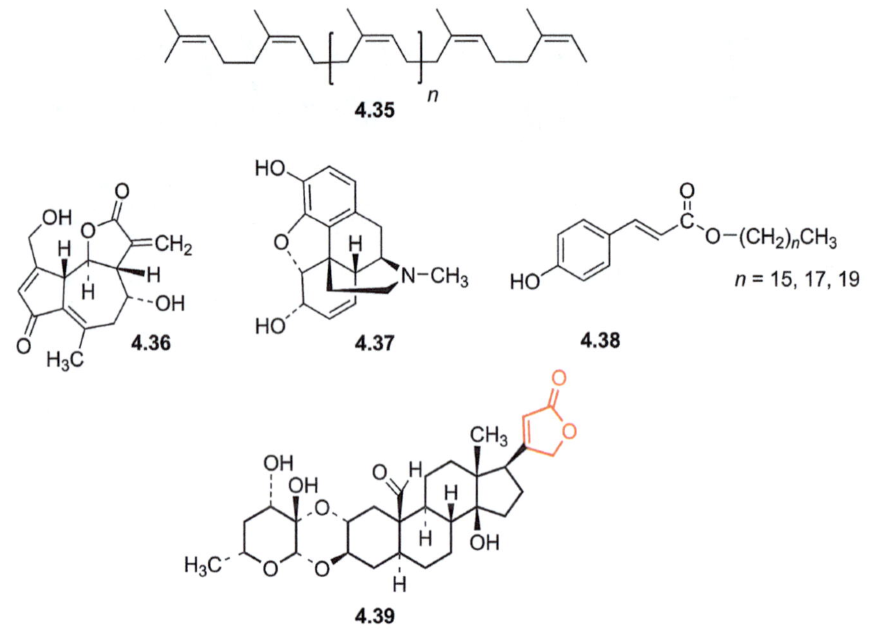

Figure 4.18 Examples of components of plant latex. Rubber **4.35** or *cis-*1,4-polyisoprene, and the sesquiterpene lactucin **4.36** are terpenoids. Morphine **4.37** is an alkaloid in the latex of opium poppy (*Papaver somniferum*). Phenolic coumarate esters **4.38** are defensive components of the latex in sweet potato plants (*Ipomoea batatas*). Calotropin **4.39** is a typical cardenolide found in milkweeds (*Asclepias* spp.). The ring structure attached to the steroid system and shown in red is characteristic of cardenolides.

it relates to the elasticity of *cis*-1,4-polyisoprene, its ability to co-agulate, and its adhesiveness to insect surfaces.

Bitter-tasting substances are often part of latex. Leaves and stems of fully grown dandelions (*Taraxacum officinale*) or lettuce heads (*Lactuca sativa*) that are bolting (*i.e.* blooming and going to seed) contain bitter terpenoids in their latex, like the sesquiterpenoid lactucin ($C_{15}H_{16}O_3$) **4.36**. Alkaloids, like morphine **4.37** from the latex of the opium poppy (*Papaver somniferum*) occur in some latex mixtures. Sweet potatoes (*Ipomoea batatas*) in the morning glory family (Convolvulaceae) are an important tropical food crop. All parts of the vines contain latex, with long-chain esters of the phenolic coumaric acid being part of it **4.38**. While the sweet potato weevil (*Cylas formicarius*) can do great damage to the crops, the insects generally avoid the tips of the vines, the plant parts with the highest concentrations of the phenolic esters.[26]

The latex of milkweeds, as well as of plants in the dogbane family and the mulberry family (Moraceae), contains cardiac glycosides. These compounds inhibit Na^+/K^+-ATPases, important for the maintenance of the electric potential in most animal cells, thus are toxic to many animals. Cardiac glycosides are steroid glycosides. Figure 4.18 shows calotropin **4.39**, a cardiac glycoside, or more specifically a cardenolide, found in milkweeds (*Asclepias* spp.). Molecules of cardenolides have a characteristic ring structure, shown in red, attached to their steroid structure. More details on cardiac glycosides and their activities will be shown in Chapter 8.2.

Latex can contain also enzymes, like the chitinases, *i.e.* enzymes that degrade chitin, the characteristic component of insects' exoskeletons.

Latex is an economical and quick-acting means of plant defense against insects. The considerable pressure under which it is stored in the laticifers has the mixture, with all its defensive compounds, rapidly moved to the site of injury in plant tissue. As latex clots rapidly, it not only damages feeding insects, but also seals plant wounds, thus keeping the liquid from oozing out further. The idea that latex protects plants from herbivores was first proposed in 1887 by Joseph F. James. He stated about latex in milkweeds that it

> "... carries with it at the same time such disagreeable properties that it becomes a better protection to the plant from enemies than all the thorns, prickles, or hairs that could be provided. In this plant, so copious and so distasteful has the sap become that it serves a most important purpose in its economy."[27]

4.9 Conclusions

This chapter has addressed the great diversity of chemical defenses of plants against insects. Examples of their formation in plants and of their actions against insects were shown. Major classes of natural products, namely terpenoids, alkaloids, steroids, and phenolics are part of the plant defenses. Some adaptations of specific insects to the plant defenses were included, to point out the ongoing coevolution of insects and plants.

What do insects obtain from plants for their survival? Which insects use plant defenses for their own protection? And how do chemical defenses of plants compare with insect defenses? Read on about these topics in Part 2 on the insect perspective.

References

1. C. M. Smith, Biochemical Plant Defenses Against Herbivores, in *All Flesh is Grass*, ed. J. Seckbach and Z. Dubinsky, Springer, Heidelberg, 2011, pp. 289–310.
2. C. C. Labandeira, Early history of arthropod and vascular plant associations, *Annu. Rev. Earth Planet. Sci.*, 1998, **26**, 329.
3. J. H. Langenheim, *Plant Resins*, Timber Press, Portland, OR, 2003.
4. http://www.fs.usda.gov/Internet/FSE_DOCUMENTS/ stelprdb5384837.pdf (accessed September 2016).
5. M. A. Phillips and R. B. Croteau, Resin-based defenses in conifers, *Trends Plant Sci.*, 1999, **4**, 184.
6. J. R. Hanson, *Chemistry in the Garden*, Royal Society of Chemistry, Cambridge, UK, 2007.
7. P. W. Paré and J. H. Tumlinson, Plant Volatiles as a Defense against Insect Herbivores, *Plant Physiol.*, 1999, **121**, 325.
8. J. Fürstenberg-Hägg, M. Zagrobelny and S. Bak, Plant Defense against Insect Herbivores, *Int. J. Mol. Sci.*, 2013, **14**, 10242.
9. A. J. K. Koo and G. A. Howe, The wound hormone jasmonate, *Phytochemistry*, 2009, **70**, 1571.
10. K. Sugimoto, K. Matsui, Y. Iijima, Y. Akakabe, S. Muramoto, R. Ozawa, M. Uefune, R. Sasaki, K. M. Alamgir, S. Akitake, T. Nobuke, I. Galis, K. Aoki, D. Shibata and J. Takabayashi, Intake and transformation to a glycoside of (Z)-3-hexenol from infested neighbors reveals a mode of plant odor reception and defense, *Proc. Natl. Acad. Sci. U. S. A.*, 2014, **111**(19), 7144.
11. M. E. Hoballah and T. C. J. Turlings, The role of fresh *versus* old leaf damage in the attraction of parasitic wasps to herbivore-induced maize volatiles, *J. Chem. Ecol.*, 2005, **31**(9), 2003.
12. A. Gfeller, M. Laloux, F. Barsics, D. E. Kati, E. Haubruge, P. du Jardin, F. J. Verheggen, G. Lognay, J.-P. Wathelet and M. L. Fauconnier, Characterization of Volatile Organic Compounds Emitted by Barley (Hordeum vulgare L.) Roots and Their Attractiveness to Wireworms, *J. Chem. Ecol.*, 2013, **39**(8), 1129.
13. K. M. Olsen and M. C. Ungerer, Freezing tolerance and cyanogenesis in white clover, *Int. J. Plant Sci.*, 2008, **169**(9), 1141.

14. M. Zagrobelny, S. Bak, A. Vinther Rasmussen, B. Jørgensen, C. M. Naumann and B. Lindberg Møller, Cyanogenic glucosides and plant-insect interactions, *Phytochemistry*, 2004, **65**, 293.
15. M. Zagrobelny and B. Lindberg Møller, Cyanogenic glucosides in the biological warfare between plants and insects: The Burnet moth-Birdsfoot trefoil model system, *Phytochemistry*, 2011, **72**, 1585.
16. R. J. Hopkins, N. M. van Dam and J. J. A. van Loon, Role of glucosinolates in insect-plant relationships and multitrophic interactions, *Annu. Rev. Entomol.*, 2009, **54**, 57.
17. J. B. Harborne, *Introduction to Ecological Biochemistry*, Academic Press, London, 4th edn, 1993.
18. H. Schmutterer, in *The Neem Tree*, ed. H. Schmutterer, VCH, Weinheim, 1995, pp. 1–34.
19. G. E. Veitch, A. Boyer and S. V. Ley, The Azadirachtin Story, *Angew. Chem., Int. Ed.*, 2008, **47**, 9402.
20. R. L. Blackman and V. F. Eastop, *Aphids on the World's Crops: An Identification and Information Guide*, Wiley, New York, 2000.
21. M. Hesse, *Alkaloids, Nature's Curse or Blessing?* Wiley-VCH, 2002.
22. T. Aniszewski, *Alkaloids – Secrets of Life: Alkaloid Chemistry, Biological Significance, Applications and Ecological Role*, Elsevier, Amsterdam, Oxford, 2007.
23. J. S. Ramsey, D. A. Elzinga, P. Sarkar, Y.-R. Xin, M. Ghanim and G. Jander, Adaptation to Nicotine Feeding in *Myzus persicae*, *J. Chem. Ecol.*, 2014, **40**, 869.
24. T. Eisner, *For Love of Insects*, Harvard University Press, 2003.
25. A. A. Agrawal and K. Konno, Latex: a Model for Understanding Mechanisms, Ecology, and Evolution of Plant Defense Against Herbivory, *Annu. Rev. Ecol., Evol., Syst.*, 2009, **40**, 311.
26. M. E. Snook, E. S. Data and S. J. Kays, Characterization and quantitation of hexadecyl, octadecyl, and eicosyl esters of p-coumaric acid in the vine and root latex of sweetpotato (*Ipomoea batatas*), *J. Agric. Food Chem.*, 1994, **42**, 2589.
27. J. F. James, The Milkweeds, *Am. Nat.*, 1887, **21**, 605.

Part 2: The Insect Perspective

5   Insects and Their Chemistry

5.1   Introduction

This chapter presents an introduction to some biological and chemical aspects of insects.

It serves as a support for readers unfamiliar with the biology and ecology of this class of animals and provides a helpful background for the later descriptions of insects interacting with plants, as seen from the insect perspective. Insects produce volatile compounds to attract mates or deter predators. Many insects appear colorful; the colors being produced by special structural features as well as by pigments. Diverse chemical compounds are used by insects as defenses. It is interesting to compare these insect compounds with volatiles, pigments, and chemical defenses produced by plants, to note similarities as well as differences, and to look at the origin of the insect chemicals. Examining the molecular structures of the insect compounds will add to the familiarity with organic compounds.

Insects are invertebrates, meaning they are animals with no spine. Most known animal species belong to the invertebrates. They include phyla of animals as diverse as mollusks (snails and squids), earthworms, starfish, and arthropods. Insects belong to the arthropods and therefore are characterized by segmented bodies and jointed legs. Insects are by far the largest class of arthropods, with about one million known species. Undoubtedly, many more will be detected. Among the insects, beetles (Coleoptera), flies (Diptera), and butterflies and moths (Lepidoptera) are the largest insect orders.[1,2] (Refer to the Glossary for more details on naming and organizing insects.)

The Chemistry of Plants and Insects: Plants, Bugs, and Molecules
By Margareta Séquin
© Margareta Séquin 2017
Published by the Royal Society of Chemistry, www.rsc.org

Insects live in almost every environment on earth: in deserts, rain forests, alpine areas, and even in Antarctica. (Oceans are the only habitat where very few insects reside.) Several factors seem to have led to the great success of this class of animals. Insects are generally small, with an average length of about 3 to 20 mm. Because of their small sizes they require little energy to survive and can hide in small niches and crevices. Figure 5.1 shows examples of insects in diverse habitats and some extremes of insect sizes. Tiny glacier fleas (*Desoria saltans*), 1.5 to 2.5 mm in length, are permanent inhabitants on alpine glaciers and snowfields and can survive temperatures down to −15 °C (Figure 5.1(a)). These insects are actually springtails (Collembola), their common name referring to their very small size. Some tropical insects, like the Costa Rican walking stick insect (order Phasmida, Figure 5.1(b)) or the neotropical giant long-horned beetle (*Titanus giganteus*), are exceptionally large, with a length of more than 20 cm. The evolution of wings enabled many insects to move to suitable environments if needed. The short life spans of insects, with rapidly changing life cycles, permit quick adaptations to changing environments.

In addition, a tough outer skeleton or *exoskeleton* protects insects. It is made of *chitin* in combination with cross-linked proteins. Chitin is a polymer related to polysaccharides (Figure 5.2). The monomers in chitin are *N*-acetylglucosamine **5.1**, a nitrogen-containing compound related to D-glucose. The monomers are connected in linear arrangements by sterically defined glycosidic linkages, namely 1,4-β-linkages. A segment of the chitin polymer **5.2** illustrates these

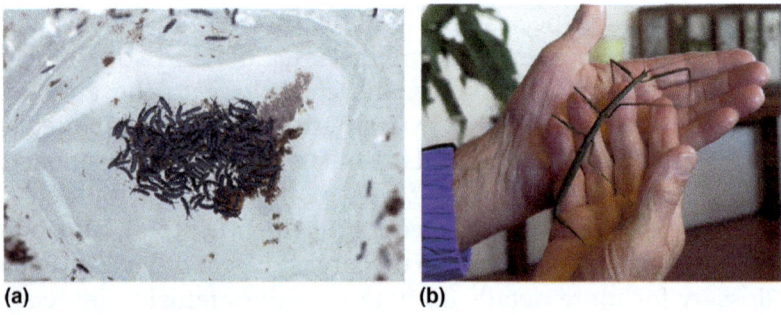

(a)   (b)

Figure 5.1 Insects in diverse environments. (a) Tiny glacier fleas or glacier springtails (*Desoria saltans*), 1.5 to 2.5 mm long, survive temperatures down to −15 °C. (Photo by Juerg Alean.) (b) The Costa Rican walking stick insect shown is one of the largest insects.

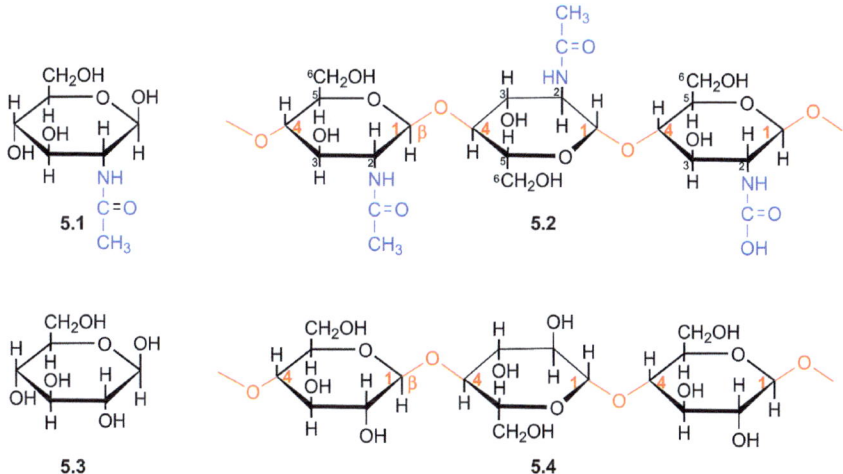

Figure 5.2 Chitin and cellulose. *N*-acetylglucosamine **5.1** is the monomer in chitin. The monomers are linked by 1,4-β-linkages in linear arrangements **5.2**. Similarly, D-glucose monomers (shown in β-configuration **5.3**) are linked by 1,4-β-linkages in cellulose **5.4**.

(a)   **(b)**

Figure 5.3 Tough exoskeletons, composed of chitin and cross-linked proteins, protect insects. (a) A ten-lined June beetle (*Polyphylla decemlineata*), with its hard, striped mantle. (b) Overwintering convergent ladybird beetles (*Hippodamia convergens*).

linkages in Figure 5.2. The rigid exoskeletons act as armor for the June beetle (*Polyphylla decemlineata*) in Figure 5.3(a) as well as for the ladybird beetles in Figure 5.3(b). (Chitin is also found in other living organisms, *e.g.* in the hard shells of crabs or lobsters, there combined with calcium carbonate in a protein/chitin matrix, and in the cell walls of fungi.) Because the exoskeletons of insects cannot expand as animals grow, insects have to go through several developmental stages (metamorphoses) in which they shed their chitin mantles and grow new, larger ones. Chitin closely resembles the structure of cellulose, the structural material of green plant leaves and stems. In cellulose, D-glucose molecules **5.3** are the monomers which are similarly connected by 1,4-β-linkages **5.4** to form the polymer. Insects and other animals (including humans) do not have the enzymes to break down these β-glycosidic linkages and cannot digest the polymers. Only animals that have microorganisms in their guts that provide the proper enzymes are capable of digesting cellulose. Chitinases, *i.e.* enzymes that break down chitin, can be found in the latex of some plants, as mentioned in Chapter 4.8.

The exterior shapes of insect bodies vary greatly. They are basically composed of three regions: head, thorax, and abdomen (Figure 5.4). The head has specific mouthparts that relate to the type of food a particular insect eats and how it is ingested. The insects' heads also have eyes or simple photoreceptors (as in caterpillars) and sensory structures that can recognize smell and taste. The thorax has one or two pairs of wings and three pairs of legs attached to it. Butterflies and many other types of insects have footpads that are taste receptors. Aside from a heart and an aorta, insects have few blood vessels. The blood of insects flows around in the body cavity. Openings in the exoskeleton allow for the uptake of oxygen from air. Digestive and reproductive organs are also in the abdomen. Insects are cold-blooded and need energy in the form of heat and food. There is little insect activity when ambient temperatures are low (as an insect collector can experience). Figure 5.3(b) shows clusters of convergent ladybird beetles (*Hippodamia convergens*) on a chilly winter day.

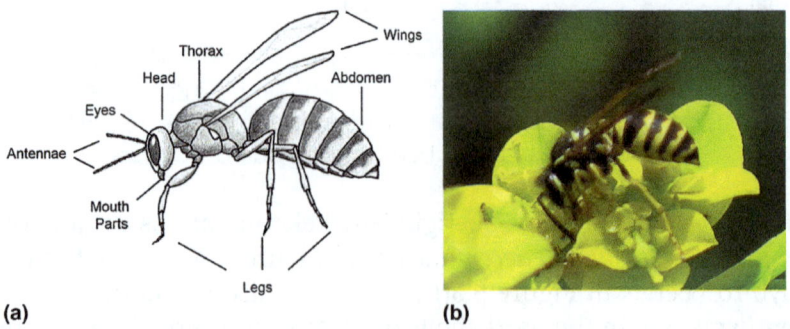

(a)　　　　　　　　　　　　　　　　　　**(b)**

Figure 5.4　Insect body parts. (a) Schematic drawing of an insect. (Drawing by Eveline Larrucea.) (b) The various body parts can be seen in this photo of a yellow jacket or wasp.

Insects go through several different developmental stages (*metamorphoses*), beginning with the egg, then several larval stages called *instars*, followed by the pupa, and ending with the adult. Metamorphoses of different insect species can include all the stages (called 'complete metamorphosis' as, *e.g.* in moths, butterflies, and beetles) or can be incomplete (as, *e.g.* in true bugs or Hemiptera; see Chapter 1). The various developmental stages of the same insect species often live in different habitats and can have dissimilar feeding habits. Think of caterpillars that eat plant leaves whereas the adult butterflies sip nectar from flowers. Many larval stages are aquatic, like mosquito larvae that eat algal and protozoan food, whereas the adults are terrestrial, the females feeding on the blood of mammals.

5.2   Insects Communicating: Pheromones

Insects release many chemical compounds that serve as a means of communication between individuals of the same insect species and that induce specific biological responses from fellow insects. These semiochemicals are generally known as *pheromones* (from Greek, meaning 'carrier of excitation'). They can act as sexual attractants, as deterrents, as trail markers, or as alarms. Insects are capable of detecting the respective compounds at very low concentrations and with high specificity.[3] Studies of the different types of pheromones are of great interest in the biological control of insects that are destructive of forest and crop plants.

Pheromones are volatile organic compounds. Some are metabolized from fatty acids by insects themselves, like the pheromone bombykol **5.5** emitted by female silkworm moths (Figure 5.5(a)). Other pheromones are metabolites from plants that insects have adapted to use as a means of communication. Insects can also modify plant volatiles to produce slightly altered structures that then serve as insect pheromones. Volatile terpenes produced by trees are commonly found in these roles as the examples below will illustrate. Pheromones are often not just one specific compound, but blends. The same volatiles may be found in different pheromones, but in different proportions, with the particular mixtures specific to an insect species. Pheromones are commonly categorized according to the behaviors they elicit in insects.

Sex pheromones are volatiles emitted by insects of one sex to attract, and sometimes to deter, individuals of the opposite sex of the same insect species. One of the best studied examples is the pheromone bombykol **5.5** (Figure 5.6); it is the volatile organic

(a)   (b)   (c)

Figure 5.5   Insects emit pheromones. (a) Silkworm moth (*Bombyx mori*). (Photo by CSIRO. Wikimedia Commons. https://commons. wikimedia.org/wiki/File:CSIRO_ScienceImage_10746_An_adult_ silkworm_moth.jpg, (accessed: October 2016).) (b) Leafcutter ants. (Photo by Wikimedia Commons. https://upload. wikimedia.org/wikipedia/commons/a/af/Leaf_cutter_ants_arp. jpg, (accessed October 2016).) (c) Signs of bark beetle activity on the trunk of a Monterey pine (*Pinus radiata*).

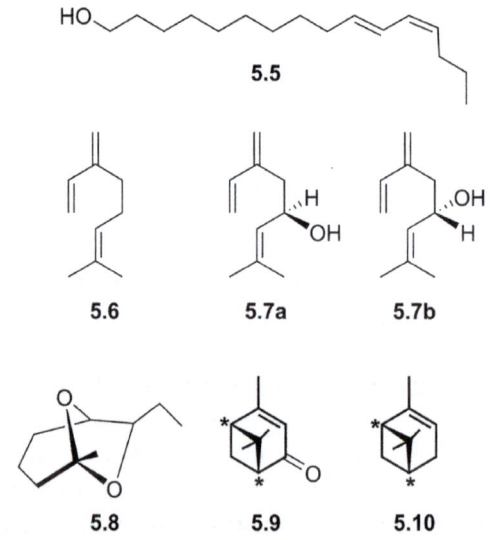

Figure 5.6   Chemical structures of pheromones. Bombykol **5.5**, also known as (*E,Z*)-10,12-hexadecadienol, is the sex pheromone of the female silk moth. Bark beetles of the genus *Ips* modify terpenes emitted by the tree bark, like myrcene **5.6**, to form ipsdienol **5.7a** or **5.7b**. *exo*-Brevicomin **5.8**, augmented by myrcene, is an aggregation pheromone of the Western pine beetle. Verbenone **5.9**, close in structure to α-pinene **5.10**, deters further beetles from approaching.

compound that female silkworm moths (*Bombyx mori*) emit to attract male mates. Bombykol was the first insect sex pheromone that was chemically characterized, in a classic study by the German chemist A. Butenandt in 1959.[4] The systematic name of bombykol is (*E,Z*)-10,12-hexadecadienol, meaning that its molecular structure consists of a 16-carbon chain with an alcohol group (OH) at one end; it also has two carbon–carbon double bonds, one in a *trans* (or *E*-) arrangement, the other one in *cis* (or *Z*-) configuration. Receptors on the antennae of male silkworm moths have been shown to be sensitive enough to detect individual molecules of bombykol in the air. Compounds similar to bombykol, with only slight structural variations, have been found as sex pheromones in other moths, like the legume pod borer (*Maruca vitrata*), a tropical, nocturnal moth that causes great damage on crops of legumes.[5]

Trail pheromones guide individuals of an insect species to a food source or to their nest. They are common in ants and termites, like the leafcutter ants in Figure 5.5(b), and also in social bees. These pheromones are formed by the insects as metabolic waste products.

Alarm pheromones are highly volatile organic compounds that are emitted by insects like bees, wasps, termites, or aphids to alert others about imminent danger.

Aggregation pheromones are released by insects in response to environmental conditions like a drought, during which weakened trees can provide food for insects. Bark beetles and their pheromones are well-researched as they contribute to major damaging infestations of pine trees (Figure 5.5(c)), among them beetles of the genus *Ips*. The structures of volatile terpenes emitted by the trees can be altered by the insects to serve as their pheromones. An example is the common monoterpene myrcene **5.6**, produced by the trees' bark, that the insects can convert into ipsdienol **5.7**, an insect aggregation pheromone. Note that ipsdienol has a chiral center. Two different species of *Ips* specifically use one of the enantiomers, **5.7a** or **5.7b**, as the attractant.[6]

Western pine beetles (*Dendroctonus brevicomis*) can destroy vast tracts of pine trees. These voracious beetles attract other individuals to the food source by releasing the volatile *exo*-brevicomin **5.8**, augmented by myrcene. The aggregation of large numbers of these beetles overwhelms the trees' defensive resins. When densities of feeding insects are getting high enough to threaten the food supply, they emit anti-aggregation or spacing pheromones. Verbenone **5.9**, close in structure to α-pinene **5.10**, deters further bark beetles from landing.[7]

5.3   Colorful Insects

The body parts of insects can be colored.[8] Think of colorful butterflies (Figure 5.7(a) and (b)) or ladybird beetles (sometimes known as ladybugs), or of an earth-colored caterpillar that blends into its environment (Figure 5.7(c)). Colors can help insects recognize and attract mates. They can also warn off potential predators, informing about the toxicity of an insect. Some shades of colorations can provide camouflage for insects (Figure 5.7(c)). Butterfly eggs are frequently brightly colored (and often have a scent mark, too), to warn off other egg-laying females – and secure an adequate food source for the emerging larvae. Insects like bees or butterflies can see a wide range of colors, including wavelengths in the ultraviolet range. Some insects perceive polarized light. Caterpillars have eye spots (ocelli) that are sensitive to close-range colors.

In contrast to most plant colors, many insect colors are the result of physical structures, like textured surfaces or scales that scatter or refract light or produce interference colors. Iridescent blues, greens, and blacks of butterfly wings are mostly formed this way. The black to blue to green iridescent wings of some butterflies, like those of the blue morpho (*Morpho* sp., Figure 5.7(a)), are brilliant enough that they can be seen from large distances. The changeable colors are created by the special arrangements of the scales on the butterflies' wings and the angles of light shining on them.

Colors in insects can also be the result of pigments in the insects' cuticles, their epidermal cells, their eyes, or in their blood. Insects

(a)   (b)   (c)

Figure 5.7   Colorful insects. (a) Microsopic scales on the back of the wings reflect light to create the iridescence of the blue morpho butterfly (*Morpho* sp.). (b) Pigments create the colors of the wings of this Gulf fritillary butterfly (*Agraulis vanillae*). (c) A caterpillar camouflaged by its earth-tone color. (Photo by Joan Hamilton.)

obtain some of the pigments by ingesting them directly from a plant source, like carotenoids (yellows or orange) or flavonoids (white to yellow). Other pigments are produced by the insects' own metabolism. Figure 5.8 shows structures of some insect pigments.

The bodies of insects are frequently surrounded by a dark cuticle. Its yellow-brown to black color is due to *melanins*. These pigments are a diverse group of polymers of highly irregular structures. Some are produced by the oxidation of the amino acid tyrosine **5.11** (Figure 5.8),

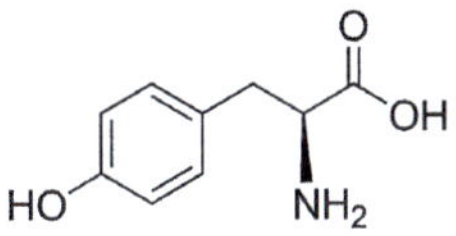

5.11

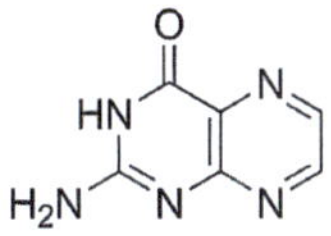

5.12

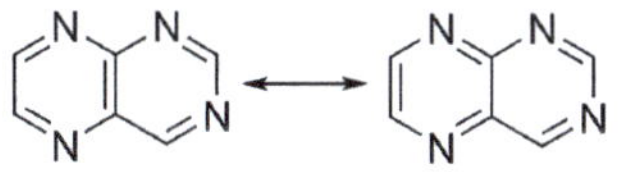

5.13

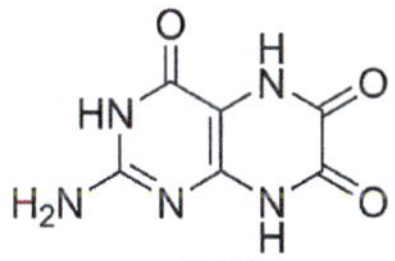

5.14

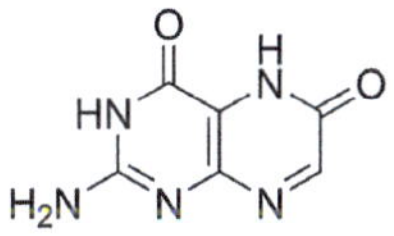

5.15

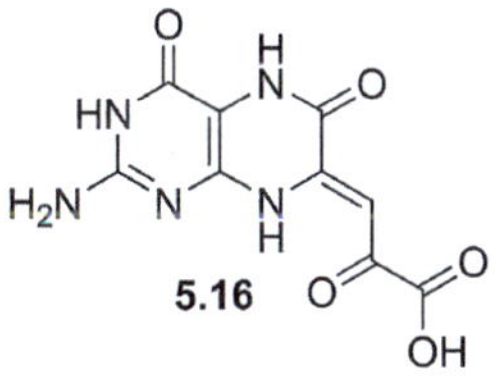

5.16

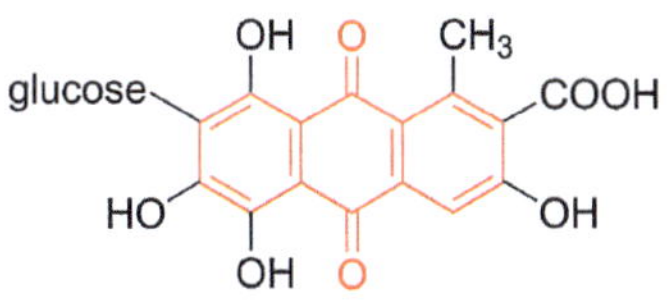

5.17

Figure 5.8 Insect pigments. Oxidized forms of the amino acid tyrosine **5.11** are monomers of some melanins. Pterins **5.12** contain the pteridine ring system **5.13** which is shown in two resonance forms. Leucopterin **5.14**, xanthopterin **5.15**, and erythropterin **5.16** are examples of pteridines found in butterfly wings. Carminic acid **5.17** is the deep-red pigment in cochineal insects. Its molecular structure contains the anthraquinone system (highlighted in red).

followed by polymerization. Different layers and thicknesses and the way they are arranged can create black, brown, yellow, or even red colors. Green insect colors are usually mixtures of pigments (but never chlorophyll), with structural colors often combining with pigments. Movement of granules that contain pigments can create changeable colorations in certain insects, as in some grasshoppers.

The first insect pigments were isolated from butterfly wings at the end of the 19th century, while their chemical structures were determined during the following decades.[9] These pigments are white, yellow, or red *pterins* **5.12** (from Greek for 'wing') that provide the respective bright colors to the insects' wings.[10] Pterin molecules have two heterocyclic rings with nitrogen atoms, called the *pteridine* ring system **5.13** (Figure 5.8). The pigments do not only absorb light within the visible spectrum. They also fluoresce in ultraviolet light, a property that contributes to the brightness of the colors. Leucopterin **5.14** is a white pigment in butterfly wings. ('Leuco' is Greek for white.) It fluoresces pale blue. Xanthopterin **5.15** is yellow and fluoresces in the yellow/green part of the spectrum. ('Xantho' is Greek for yellow.) Erythropterin **5.16** creates red, orange, and yellow colors on butterfly wings and has an orange fluorescence. ('Erythro' is Greek for red.)

Note the sequences of conjugated double bonds in the molecular structures **5.12–5.16** (as well as in **5.17**). We encountered similar sequences earlier in colorful plant pigments. They characterize the structures of organic molecules that strongly absorb light, with longer sequences of conjugated double bonds creating more colorful compounds as the wavelengths of absorbed light shift into the visible range. Compare also the minor structural differences among the shown examples of pterins. They result in different absorbed wavelengths and thus produce different colors. While the positions of the double bonds are shown as fixed in the structures, the electrons in the double bonds are in fact distributed over several bonds. More than one correct molecular structure of the pigments could be drawn, different only in the position of their electrons, as demonstrated with the example of pteridine **5.13**. The phenomenon of *resonance* of these conjugated systems creates strong absorption of light and is common in the molecular structures of colorful organic pigments in general.

While pteridines were first discovered in butterfly wings, they have since been found to be widespread in animals and have important roles in mammals as well. The pteridine structure is part of folic acid and the folates, which are vitamins of the B group. Pteridines are growth-accelerating and growth-determining agents and important in cell division.

Ommochromes are a diverse group of pigments in the eyes of insects where they determine the eyes' colors.

An unusual insect pigment is the bright-red pigment carminic acid **5.17** from cochineal insects (*Dactylopius coccus*). Carminic acid has an anthraquinone structure (highlighted in **5.17**). Note the many conjugated double bonds in the molecule. Carminic acid is likely to have a defensive function in the insects. Carminic acid and the related structure of carmine are well-known pigments among artists and dyers (see Chapter 10.6). Some bright-yellow aphids have pigments with similar anthraquinone structures.

5.4 Light-producing Insects

During a warm summer night, insects with telling names like 'glow worms', 'fire flies', or 'lightning bugs' may be observed to emit flashes of light.[11] These insects are types of beetles that, with defined sequences of light emission, attract mates or lure prey. The phenomenon is known as *bioluminescence*. The complex mechanism has been studied in detail in eastern fireflies (*Photinus pyralis*), the most common fireflies in North America. Their bioluminescent compound is luciferin **5.18** (Figure 5.9). The enzyme luciferase, stored in adjacent body cells in the insects, induces oxidation of luciferin in the presence of molecular oxygen and the energy source adenosine triphosphate (ATP). The reaction produces molecules in an electronically excited state. When electrons drop back to the ground state, the process releases energy in the form of light. The insects seem to be able to control the supply of oxygen to the reaction mixture and thus produce brief, rhythmic flashes of light that serve as signals to mates. Luciferin is a heterocyclic compound, with nitrogen and sulfur atoms in its rings and conjugated double bonds.[12]

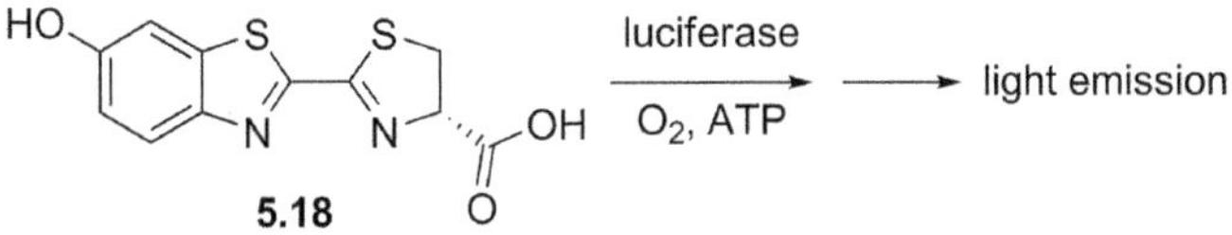

Figure 5.9 Bioluminescence in insects. Luciferin **5.18** is the bioluminescent compound in the eastern firefly (*Photinus pyralis*).

5.5   Defensive Insects

Defensive chemical compounds in insects are widespread and highly diverse. We may be familiar with the sharp scent of formic acid produced by an ant in defense or the foul smell emitted by a stink bug (Figure 5.10(a)). We are likely to have experienced the painful stings by insects like ants or bees that contain strong irritants. The descriptions below provide an introduction to some of the foul-smelling, bad-tasting, or downright toxic chemicals that insects use to defend themselves. The compounds range from simple molecules like hydrogen cyanide or formic acid **5.19** to compounds with complex molecular structures (Figure 5.11). We'll encounter chemical families described earlier, like hydrocarbons and terpenes. As heterotrophs, insects need to obtain the basic nutrients from a plant source (or by feeding on animals that fed on plants) to synthesize the chemical defenses. In addition, some insects directly use defensive plant compounds, most of them with complex chemical structures, and apply them in their own defense. (This will be addressed in Chapter 8.)

Many insects release malodorous, sometimes irritating, scents when bothered, like grasshoppers, cockroaches, and stink bugs (hence the name). Insects that give off such chemical defenses are frequently brightly colored or have striking patterns that alert potential predators to their unpalatability or toxicity. These colors are known as *aposematic* colors. We can see them, *e.g.* in the insect examples in Figure 5.10(a) and (c), also in the ladybird beetles shown earlier in Figure 5.3(b).

(a)   (b)   (c)

Figure 5.10   Insects that emit defensive chemicals. (a) Stink bug (*Banasa* sp., Pentatomidae). (Photo by Bob Case.) (b) Black blister beetles (*Epicauta* sp., Meloidae). (Photo by Steve Edwards.) (c) Burnet moths (*Zygaena* sp., Zygaenidae) release hydrogen cyanide. (Photo by Ruth Marent.)

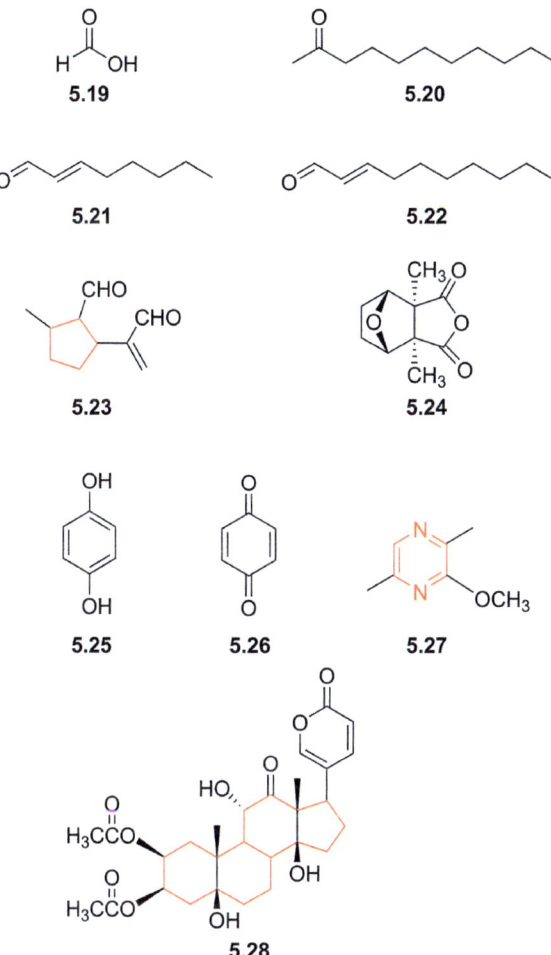

Figure 5.11 Structures of some chemical insect defenses. Formic acid **5.19**, undecanone **5.20**, *trans*-2-octenal **5.21**, *trans*-2-decenal **5.22** are relatively simple organic compounds. The structures of anisomorphal **5.23**, an iridoid, and cantharidin **5.24** are both related to terpenes. (The cyclopentane ring typical for iridoids is shown in red in **5.23**.) *p*-Hydroquinone **5.25** and its sharp-smelling oxidation product benzoquinone **5.26** are produced by bombardier beetles. An example of a methoxypyrazine **5.27** (its pyrazine ring shown in red) contributes to the chemical defense of ladybird beetles. Lucibufagin **5.28** is a steroid. (The steroid ring system is shown in red.)

Figure 5.11 shows some structures of chemical defenses that insects have evolved. They illustrate the great diversity of the chemistry of these compounds. Examples of insects and their defensive reactions including emission of stinking or irritating chemicals are vividly described in T. Eisner's book "For Love of Insects".[13]

First we address some defensive compounds with relatively simple molecular structures.

Formic acid **5.19** is an irritant with a sharp odor found in the sting of such insects as ants (some of the genus *Formica*, hence the name of the acid), bees, and wasps. It is also part of the irritating mixture of oakleaf moths (*Lochmaeus manteo*, formerly *Heterocampa manteo*) that feed on leaves of deciduous trees, especially oaks. When pestered, its caterpillar larva sprays an irritating mixture of formic acid and long-chain ketones, like undecanone **5.20**.[14]

Compounds with longer aliphatic carbon chains are also part of the malodorous oils emitted by stink bugs (Family: Pentatomidae) (Figure 5.10(a)). These insects are relatively harmless bugs that feed mostly on plants. When irritated they give off liquid mixtures with unpleasant smells (described as waxy or nutty) that contain *trans*-2-octenal **5.21**, an aldehyde with eight carbon atoms, and *trans*-2-decenal **5.22**, with ten carbons (Figure 5.11). The long carbon chains of the compounds are derived from fatty acids from the insects' metabolism.

Some structures of chemical insect defenses are related to monoterpenes, namely the so-called *iridoids* that were first isolated from a type of ant of the genus *Iridomyrmex* (hence their name). They are widespread as defenses in insects and have been found in ants and beetles. Their monoterpene structures have a cyclopentane ring (highlighted in **5.23**, Figure 5.11). Some tropical walking stick insects (*Anisomorpha buprestoides*) from southeastern North America produce irritating sprays containing anisomorphal **5.23**, an iridoid. (Try to find the two isoprene units in **5.23**.) A more complex monoterpene structure is found in cantharidin **5.24**. This compound is a strong skin irritant and *vesicant* that is part of the defensive secretions emitted by beetles with the telltale name 'blister beetles', members of the worldwide insect family Meloidae (Figure 5.10(b)). If blister beetles are ingested with animal feed like alfalfa, cantharidin is toxic enough to cause severe poisoning of horses and other animals.

The name 'bombardier beetle' sounds ominous. Indeed, when these beetles are disturbed they give off a hot mixture of *p*-hydroquinone **5.25** (Figure 5.11) and hydrogen peroxide (H_2O_2). The chemical reaction between these reactants forms benzoquinone **5.26**.

All compounds in the mixture have a characteristic sharp smell. Bombardier beetles belong to the Carabidae family and are found on all continents (except Antarctica).

Well-known and well-liked ladybird beetles have chemical defenses, too, in the form of malodorous methoxypyrazines, which they give off when threatened. An example of a methoxypyrazine is shown in structure **5.27**. Pyrazines are heterocyclic aromatic compounds with two nitrogen atoms in the ring structure (highlighted). The striking colorations and color patterns of the beetles alert potential predators. Equally brightly colored are the common burnet moths (*Zygaena* sp., Figure 5.10(c)); they release toxic hydrogen cyanide (HCN) when annoyed.

We encountered fireflies of the genus *Photinus* in Chapter 5.4 on bioluminescence. These insects also feature complex toxic steroids known as 'lucibufagins' (see example **5.28** of a lucibufagin) and are distasteful to birds. The name of the compounds is based on their close structural relationship with toxins found in toads (*Bufo* spp.). In an intricate example of acquiring chemical defense, female fireflies of the genus *Photuris* catch *Photinus* males by faking the flash signals of female *Photinus* fireflies. By eating the captured insects, the *Photuris* predators not only acquire nutrients from their prey, but also toxic lucibufagins.[15] In Chapter 8 we'll encounter similarly-structured plant steroids; they are cardiotonic (heart active) compounds that are ingested by certain insects for their own defense.

5.6 Insects and Their Hormones

Hormones regulate many physiological processes in the bodies of insects. The chemical compounds that compose hormones are synthesized by the insects and transported by way of their body fluids to the sites in the insect system where needed. Extremely small quantities of these compounds can induce long-term developmental changes. Examples of insect hormones include those that regulate molting processes and all the stages of metamorphosis, as well as sexual behavior, reproduction, and feeding behavior of an insect.[16] (Note the difference with pheromones, chemical compounds that stimulate specific behavior in other individuals.)

There are three major classes of insect hormones: the neurohormones, the juvenile hormones, and the molting hormones. The largest class consists of the neurohormones that regulate diverse functions in the insect body. Chemically they are peptides and proteins.

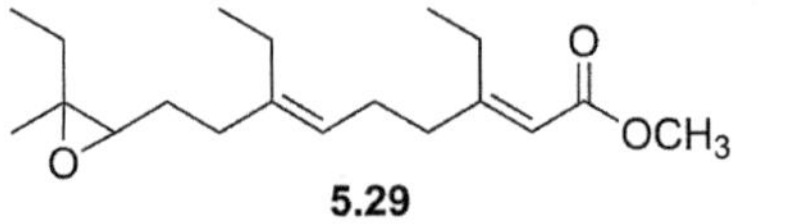

Figure 5.12   Insect hormones. Juvenile hormone 0 (JH 0) **5.29** is one of
the sesquiterpene hormones that control insect development.
Ecdysone **5.30** is a steroid that is required to induce molting in
insects.

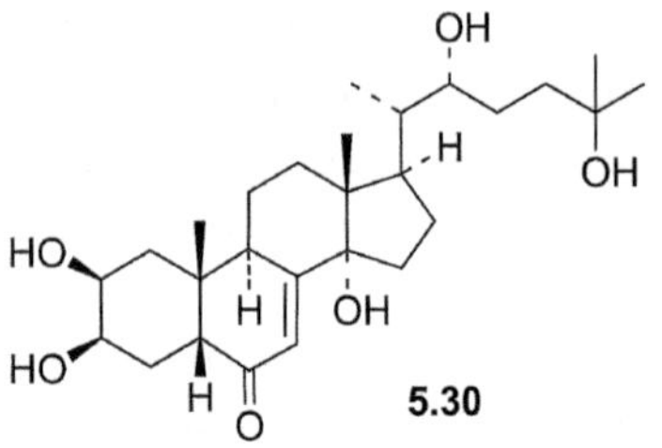

Our focus here is on the classes of the juvenile hormones, commonly
abbreviated as JH, and on the molting hormones or ecdysteroids. These
two classes have been the focus of classical chemical research, in part in
attempts to find methods to control insect pests.[17,18]

Juvenile hormones are a group of closely related acyclic sesquiterpenes – with fifteen carbon atoms (Figure 5.12). They control
metamorphosis and regulate the reproductive development. Insects
biosynthesize these hormones from fatty acids in their bodies.
An example is juvenile hormone 0 (JH 0) **5.29**. Molting of the insects
requires ecdysteroid hormones, with ecdysone **5.30** found in most
immature insects. Note the typical steroid structure. Insects cannot
synthesize steroids *de novo*. Therefore, steroids, like cholesterol, must
be part of their nutrition (see the following chapter on insect diets).

A note on insect hormones in comparison with so-called plant
hormones or phytohormones (like ethylene) is in order. Phytohormones also act as signaling compounds in minute amounts and
regulate plant growth. But unlike animal hormones they are not
produced in a central plant organ; rather, each plant cell is able to
produce the substances. Therefore, a more adequate expression that
avoids confusion is 'plant growth substances'.

5.7   Conclusions

The structures described in this chapter show the great diversity of
organic compounds that insects need to function as individuals, to
defend themselves, and to find mates. Much research has been conducted to elucidate the structures of the compounds, and research
continues. These studies contribute to the understanding of insect
systems and how insects interact with other individuals.

Read on about the nutrients that insects need to ingest to be able to compose these vital compounds and how herbivorous insects obtain them.

References

1. R. J. Elzinga, *Fundamentals of Entomology*, Pearson, Upper Saddle River, NJ, 6th edn, 2004.
2. P. J. Gullan and P. S. Cranston, *The Insects: An Outline of Entomology*, Wiley-Blackwell, Chichester, West Sussex, UK, 4th edn, 2010.
3. R. F. Chapman, J. N. McNeil and J. G. Millar, Chemical communication: pheromones and allelochemicals, in *The Insects: Structure and Function*, ed. S. J. Simpson and A. E. Douglas, Cambridge University Press, Cambridge, 5th edn, 2013, ch. 27.
4. A. Butenandt, R. Beckmann and E. Hecker, Über den Sexuallockstoff des Seidenspinners.1. 'Der biologische Test und die Isolierung des reinen Sexuallockstoffes Bombykol, *Hoppe-Seyler's Z. Physiol. Chem.*, 1961, **324**(1), 71.
5. S. Schläger, F. Beran, A. T. Groot, C. Ulrichs, D. Veit, C. Paetz, B. R. M. Karumuru, R. Srinivasan, M. Schreiner and I. Mewis, Pheromone Blend Analysis and Cross-Attraction among Populations of *Maruca vitrata* from Asia and West Africa, *J. Chem. Ecol.*, 2003, **29**(4), 989.
6. D. R. Miller, J. H. Borden, G. G. S. King and K. N. Slessor, Ipsenol: an aggregation pheromone for *Ips latidens*, *J. Chem. Ecol.*, 1991, **17**(8), 1517.
7. J. A. Byers, D. L. Wood, J. Craig and L. B. Hendry, Attractive and inhibitory pheromones produced in the bark beetle, *Dendroctonus brevicomis*, during host colonization, *J. Chem. Ecol.*, 1984, **10**(6), 861.
8. R. F. Chapman, P. Vukusic and L. Chittka, Visual signals: color and light production, in *The Insects: Structure and Function*, ed. S. J. Simpson and A. E. Douglas, Cambridge University Press, Cambridge, 5th edn, 2013, ch. 25.1–25.6.
9. W. Pfleiderer, Recent Developments in the Chemistry of Pteridines, *Angew. Chem., Int. Ed.*, 1964, 3(2), 114.
10. B. Wijnen, H. L. Leertouwer and D. G. Stavenga, Colors and pigmentation of pierid butterfly wings, *J. Insect Physiol.*, 2007, **53**(12), 1206.

11. R. F. Chapman, P. Vukusic and L. Chittka, Visual signals: color and light production, in *The Insects: Structure and Function*, ed. S. J. Simpson and A. E. Douglas, Cambridge University Press, Cambridge, 5th edn, 2013, ch. 25.7.

12. E. H. White, F. McCapra, G. F. Field and W. D. McElroy, The structure and synthesis of firefly luciferin, *J. Am. Chem. Soc.*, 1961, **83**(10), 2402.

13. T. Eisner, *For Love of Insects*, Harvard University Press, 2003.

14. T. Eisner, A. F. Kluge, J. C. Carrel and J. Meinwald, Defense mechanisms of arthropods. XXXIV. Formic acid and acyclic ketones in the spray of a caterpillar, *Ann. Entomol. Soc. Am.*, 1972, **65**, 765.

15. T. Eisner, M. A. Goetz, D. E. Hill, S. R. Smedley and J. Meinwald, Firefly "femmes fatales" acquire defensive steroids (lucibufagins) from their firefly prey, *Proc. Natl. Acad. Sci. USA*, 1997, **94**(18), 9723.

16. R. F. Chapman and S. Reynolds, Endocrine system, in *The Insects: Structure and Function*, ed. S. J. Simpson and A. E. Douglas, Cambridge University Press, Cambridge, 5th edn, 2013, ch. 21.

17. H. Röller, K. H. Dahm, C. C. Sweeley and B. M. Trost, *Angew. Chem. Internat. Edition*, 1967, **6**, 179.

18. K. Nakanishi, The ecdysones, *Pure Appl. Chem.*, 1971, **25**(1), 167.

6 Insects Feeding on Plants

6.1 Introduction

About half of all the insect species are *phytophagous*, which means that they eat plants or plant materials. They consume leaves, stems, and roots, feed on pollen or nectar, or eat wood from trees. Some insects, like many grasshoppers, locusts, and certain aphids, are generalists (*polyphagous*); they obtain their food from numerous different plants. For example, the caterpillar larvae of gypsy moths (*Limantria dispar dispar*, Figure 6.1(a)) are notorious for their generalist feeding behavior; they infest more than five hundred different plants.[1] Other insects, known as *oligophagous*, limit their feeding to several related plants. Examples are the cabbage white butterflies (*Pieris brassicae*, Figure 4.10(b)) that select plants of the cabbage family (Brassicaceae), or larvae of borer beetles (*Ergates spiculatus*) tunneling into the heartwood of different conifers (Figure 6.1(b)). The caterpillars of anise swallowtail butterflies (*Papilio zelicaon*, Figure 6.1(c)) consume mostly plants of the carrot family (Apiaceae), like fennel plants (*Foeniculum vulgare*). Some types of insects are *monophagous*, feeding only on one type of plant. For example, alfalfa aphids, as the common name implies, limit their feeding to alfalfa (*Medicago sativa*), a legume. Oligo- and monophagous insects often obtain chemical defenses from their host plants. This will be the special topic of Chapter 8.

The mouthparts at the insects' heads determine what type of food a particular insect can ingest. Mouthparts can be piercing-sucking, as in aphids, or chewing-lapping, as in bees, or simply chewing, as in beetles and caterpillars. There are also insect mouthparts adapted for

The Chemistry of Plants and Insects: Plants, Bugs, and Molecules
By Margareta Séquin
© Margareta Séquin 2017
Published by the Royal Society of Chemistry, www.rsc.org

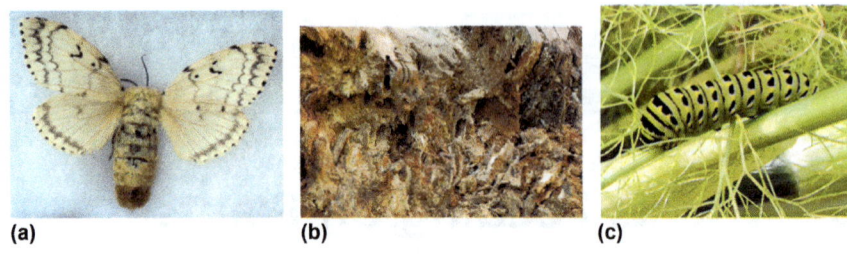

(a)   (b)   (c)

Figure 6.1   Examples of polyphagous and oligophagous insects. (a) A
female gypsy moth (*Limantria dispar dispar*), the adult of the
polyphagous caterpillar that infests vast numbers of many
different plants. (b) Damage by larvae of oligophagous borer
beetles (*Ergates spiculatus*) tunneling in ponderosa pine wood.
(Photo by Anne Yniguez.) (c) Caterpillars of the anise swallowtail
(*Papilio zelicaon*) preferably feed on fennel plants (*Foeniculum
vulgare*) and other plants of the Apiaceae. (Photo by Alex
Madonik.)

mining leaves or boring channels in wood. The structures of the insects' guts determine what they can digest. Because the patterns and
traces of plant damage can be related to the feeding insects, they can
be used not only to track down infesting insects today, but also to
determine the types of ancient phytophagous insects that caused
damage in fossil plants (Figure 4.1(b)).

6.2   Basic Insect Diet

The dietary requirements of insects are comparable to those of other
animals in many ways – and then again, there are distinct differences.
Insects, as heterotrophs, have to ingest simple sugars, amino acids,
and vitamins. They also need water and minerals. On the other hand,
they can synthesize fatty acids in their bodies (with a few exceptions).
But all insects need to have a nutritional source of cholesterol, which
is required as a component of their cell membranes and for the
synthesis of their steroid hormones. Unlike other animals (including
humans), insects cannot synthesize cholesterol. Phytophagous
insects have to obtain the required nutrients from living or decomposing plants.

A common method to learn about their nutritional needs is to feed
artificial diets to insects in a laboratory. Aphids, the topic of numerous investigations because of their infestations of crop plants, are

commonly raised on synthetic media in studies about their feeding requirements and preferences. Table 6.1 shows a sample diet solution commonly used to raise aphids and other insects.[2] It is an aqueous solution that contains amino acids, sucrose, vitamins, and trace metals, as required by the insects. Artificial diets are conveniently available for year-round studies as they are prepared from commercial ingredients. The development of suitable diet solutions is challenging and time-consuming. But once their compositions are optimized, they are simple to prepare and easy to use. Additional compounds, like plant secondary compounds, can be added to the basic diet solutions to study insects' feeding behavior towards them.

Note the list of L-amino acids, required for the synthesis of proteins, in the sample diet solution in Table 6.1. It includes the *essential amino acids*, *i.e.* amino acids that insects cannot synthesize in their systems. They are: L-histidine, L-isoleucine, L-leucine, L-methionine, L-phenylalanine **2.23**, L-threonine, L-tryptophan, and L-valine. The diet solution also contains non-essential amino acids, like L-proline **2.24**, which insects can synthesize but not in sufficient quantities. Insect species vary in their particular nutritional needs.

Figure 6.2 shows structures of nutrients that insects must ingest with their nutrition. This includes a dietary source of vitamins, *i.e.* organic compounds that are required by an organism in small quantities and that are essential for normal growth. Vitamins are commonly divided into water-soluble and fat-soluble ones. Water-soluble vitamins must be a part of the insects' diet, just as for most other animals (including humans). They include vitamins of the B group, like thiamine (B_1) **6.1**, riboflavin (B_2) **6.2**, pyridoxine (B_6) **6.3**, nicotinic acid (B_3) **6.4**, pantothenic acid (B_5) **6.5**, as well as ascorbic acid (vitamin C) **6.6**, and biotin (vitamin H) **6.7**. Note the many polar functional groups, like COOH and OH groups, in compounds **6.1**–**6.7** that indicate their water-solubility. Notice also the numerous chiral centers which must be correctly oriented for proper biological functioning. Unlike the water-soluble vitamins, insects require only minimal amounts of some fat-soluble vitamins, like β-carotene **2.31** of the vitamin A group, and calciferol **6.8** of the vitamin D group. Their requirement became evident only when insects showed developmental deficiencies after many insect generations were raised on diets without these vitamins. As for other fat-soluble vitamins, insects do have the ability to synthesize their own vitamins E and K.[3] Compare these requirements with the components of the synthetic insect diet in Table 6.1.

Table 6.1 Composition of a synthetic diet medium used to raise pea aphids (*Acyrthosiphon pisum*), adapted from Akey and Beck (see reference in text).[a]

Amino acids	mg	Vitamins	mg	Trace metals	mg	Other	Amount
L-Alanine	100	p-Aminobenzoic acid	10.0	Cupric chloride	0.245	Calcium citrate	10.0 mg
L-Arginine HCl	400	L-Ascorbic acid	100.0	Ferric chloride	1.336	Cholesterol benzoate	2.5 mg
L-Asparagine	300	Biotin	0.1	Manganese chloride	0.504	Magnesium chloride hexahydrate	200.0 mg
L-Aspartic acid	100	D-Calcium pantothenate	5.0	Sodium chloride	1.271	Potassium dihydrogen phosphate	250.0 mg
L-Cysteine HCl monohydrate	50	Choline chloride	50.0	Zinc chloride	0.417	Sucrose	35.0 g
L-Cystine	5	Folic acid	1.0				
γ-Aminobutyric acid	20	meso-Inositol dihydrate	50.0				
L-Glutamic acid	200	Nicotinic acid	10.0				
L-Glutamine	600	Pyridoxine hydrochloride	2.5				
Glycine	20	Riboflavin	5.0				
L-Histidine	200	Thiamine hydrochloride	2.5				
DL-Homoserine	800						
L-Isoleucine	200						
L-Leucine	200						
L-Lysine HCl	200						
L-Methionine	100						
L-Phenylalanine	100						
L-Proline	100						
L-Serine	100						
L-Threonine	200						
L-Tryptophane	100						
L-Tyrosine	20						
L-Valine	200						

[a]Formulation: pH adjusted to 7.5 with KOH 1.75 M and distilled-deionized water to make 100 ml of diet solution.

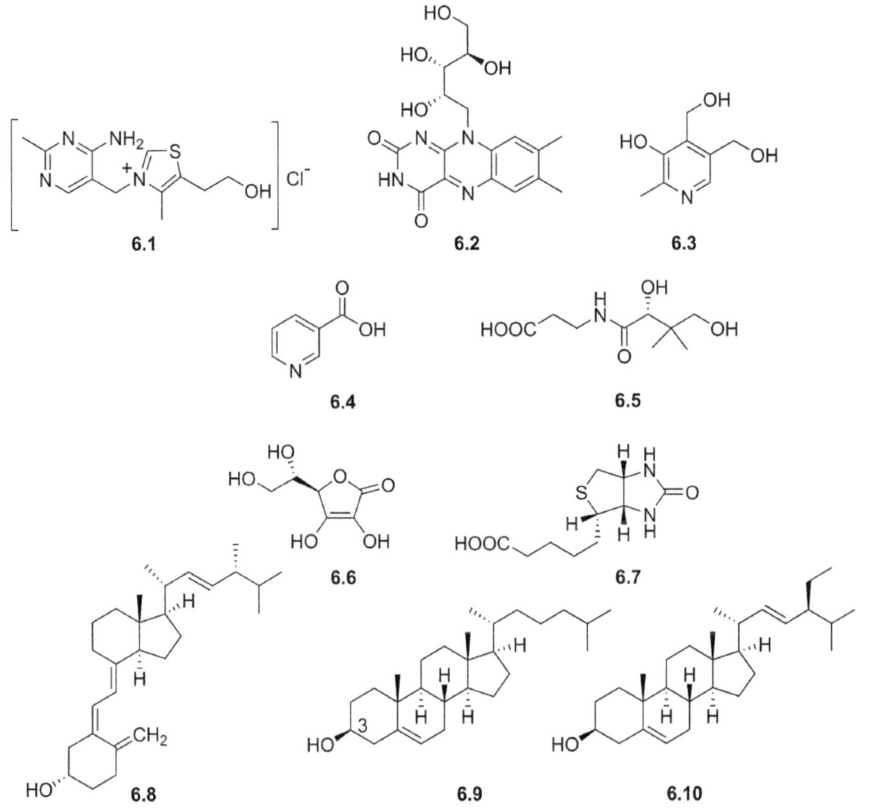

Figure 6.2 Required nutrients for insects. Insects need to ingest the water-soluble vitamins thiamine (B_1) **6.1**, riboflavin (B_2) **6.2**, pyridoxine (B_6) **6.3**, nicotinic acid (B_3) **6.4**, pantothenic acid (B_5) **6.5**, as well as ascorbic acid (vitamin C) **6.6**, and biotin (vitamin H) **6.7**. They require very small amounts of calciferol **6.8**. Insects must consume sterols like cholesterol **6.9**, or stigmasterol **6.10**, a plant steroid.

Unlike most vertebrates, insects lack enzymes to synthesize steroids like cholesterol **6.9** (Figure 6.2). The OH functional group in the 3 position (pointed out in **6.9**) assigns cholesterol to the subgroup of 'sterols'. It is a compound that is not found in plants. But low concentrations of other sterols, the 'phytosterols', do occur in many plants. An example is stigmasterol **6.10**, found in soybeans and many other legume plants (Fabaceae). Note the structural similarity of stigmasterol compared to cholesterol. Insects have the enzymes to transform plant sterols into the required cholesterol. They are then able to synthesize hormones like ecdysteroids and growth factors.

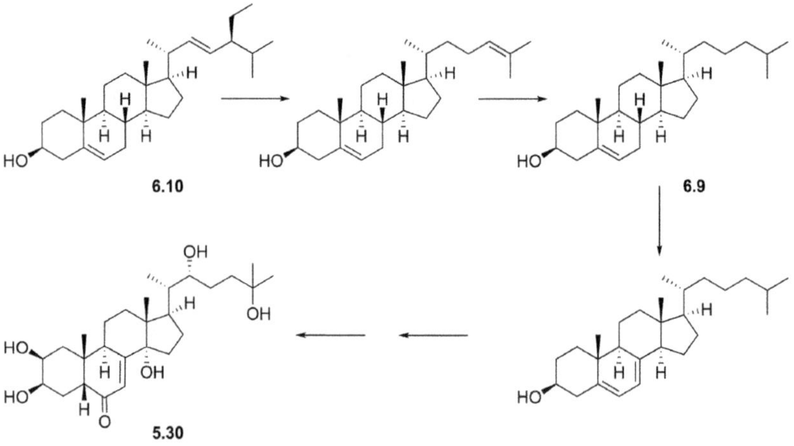

Figure 6.3 Simplified scheme of metabolic pathways leading from phytosterols, like stigmasterol **6.10**, to cholesterol **6.9** and to insect hormones like ecdysone **5.30**.

Figure 6.3 shows a simplified scheme of the metabolic pathways that lead from phytosterols, like stigmasterol, to cholesterol **6.9**, and from there to molting hormones like ecdysone **5.30**. Insects that feed on sterol-deficient diets generally grow to immature stages, but fail to develop into adults. (The synthetic medium shown in Table 6.1 contains cholesterol benzoate, a derivative of cholesterol.) Note also that the non-polar structures of β-carotene **2.31**, calciferol **6.8**, cholesterol **6.9**, and stigmasterol **6.10**, consisting mostly of hydrocarbon structures, point to their lipophilic (and non-water-soluble) properties.

Most terrestrial insects are highly adapted for water conservation and get most of the water they need directly from their food. Some insects, especially in deserts, obtain the little water they require from drops of dew.

How do plant-eating insects acquire the required nutrients, and in sufficient amounts? Plants are not an ideal source of them because they do not supply cholesterol and contain only low amounts of phytosterols. In addition, insects require a nutrition high in nitrogen and phosphorus content, something that plants provide only poorly. (Insects themselves have a higher nitrogen content than plants – which makes insects a good protein source as food for higher animals.) Therefore, in order to obtain adequate amounts of the required nutrients, insects need to eat large quantities of plant material with respect to their body weights. They generally seek out the most nutritious plants and prefer those with a high nitrogen content, like plants of the pea family (Fabaceae). (Figure 6.4(a)).

(a)   (b)   (c)

Figure 6.4 Phytophagous insects obtaining their required nutrients. (a) Aphids feeding on the phloem of a lupine (Fabaceae). (b) Blooming coconut palm, with wasp pollinators feeding on pollen that provides them with proteins and starch. (c) Butterflies obtaining minerals from the soil.

Chewing insects, like beetles and caterpillars, ingest the plants' cellulose, which is indigestible for them, and obtain nutrients from all the plant parts. This includes the *xylem*, which is the vascular tissue through which most of the water and minerals of a plant are conducted, and also the *phloem*, the part of vascular plant tissues that transports the sugars and has all the amino acids required by insects. The nitrogen content of plants' xylem is very low, about ten times lower than that of the phloem. Therefore, the phloem is much more nutritious for insects. Many phytophagous insects obtain their nutrients directly from the phloem. Famous phloem-feeders are aphids that have sucking mouthparts, called 'stylets', which they can insert with precision between the plant cells and suck up the phloem solution (Figure 6.4(a)). The phloem, though, not only transports essential nutrients, but often also contains plant toxins depending on the plant. This can present problems for insects that do not have mechanisms to cope with the plant defenses.

Honey bees, and some types of wasps (Figure 6.4(b)) and beetles are pollen feeders, pollen being rich in proteins and starch (compare Chapter 2.10). Honey bees also feed on flower nectar. Adult butterflies obtain their nutrients mostly from the aqueous solutions of flower nectars (see Chapter 2.8), in addition to the occasional intake of minerals from sources like soil (Figure 6.4(c)). Some insects, like ants, feed on *honeydew*, a sugary solution excreted by insects like aphids. Honeydew tends to be lower in plant toxins and thus is better tolerated by feeding insects than phloem.

6.3   How Insects Select Food Plants

Structural plant features can influence whether insects will consider feeding on a plant part or not. Obstacles include tough leaves, bristles, fine hair, or spines on stems, or too smooth a plant surface. The chemical constituents of a plant, including their olfactory and taste characteristics, ultimately determine if insects will start consuming a plant or avoid it. Visual cues like colors influence the selection of a host plant as well. Secondary metabolites, like those affecting taste and smell, are common reasons that many insects, namely the oligophagous insects, have a preference for certain plants. Figure 6.1(c) shows the caterpillar of the anise swallowtail specializing in feeding on plants of the Apiaceae. Cabbage white butterflies mostly eat plants of the cabbage family (Brassicaceae), taking advantage of the plants' glucosinolates (Figure 4.10(b)). Many insects have adapted to feeding on plants that contain toxins (often with a particular taste) and ingest them for their own protection, as will be addressed in detail in Chapter 8. The chemical compositions of plants tend to change with location, growth rate, and the age of the plants. Tougher leaves and a higher content of defensive plant compounds are common in more mature plants and make them less attractive to infesting insects.[4,5]

The strong correlations between plant secondary products and the particular insects feeding on them are the reason that many insect guide books mention the host plants or even feature an index of the specific plants that insects seek out.[6] Users of these guides are thus encouraged to search for the food plants, to increase the likelihood of finding the respective insects.

Several techniques have been developed to study the attraction or deterrence of host plants and their components towards distinct insects, aside from mere observations. *Electroantennograms* are commonly used to detect a positive or negative response of a type of insect towards a plant and its chemical constituents. The antennae of insects are the major sites where insects receive olfactory stimuli, *i.e.* odor and taste. An electrode attached to the insects' antennae can measure the responses to particular stimuli which are recorded in an electroantennogram.[7] Artificial diet solutions have also been used to detect insect responses towards select secondary plant products. In some studies, various defensive compounds from plants shown to be resistant to certain insects were added to basic diet media. The tolerance or resistance by insects, *e.g.* aphids, indicated the feeding preferences towards natural products from specific plants.[8]

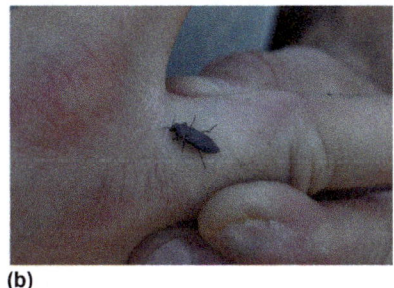

(a)   (b)

Figure 6.5 Heat sensitive insects find plant food. (a) Shrubs and trees charred by a wildfire in Northern California. (b) A charcoal beetle or fire beetle (*Melanophila acuminata*) can detect areas burnt by wildfires based on the insect's infrared sensitivity. (Photo by Joan Hamilton.)

A highly unusual example of how insects detect food plants is found in wildfire areas, *e.g.* of southwestern North America (Figure 6.5(a)). Charcoal beetles or fire beetles (*Melanophila acuminata*, (Figure 6.5(b)) are common in temperate zones of the Northern Hemisphere, including large areas of Europe and North America. These beetles are not only highly sensitive to volatile compounds typical of burning wood but also have special receptors for infrared radiation from the heat produced by wildfires. Forest fires emit smoke, sound, visible light, and infrared radiation. A typical forest fire can reach temperatures of 500 °C to more than 1000 °C. The extreme sensitivity towards infrared radiation from the fire and to heat allows the beetles to detect forest fires from large distances away; detection from more than a hundred kilometers has been reported. Huge numbers of the beetles are attracted to the wildfire areas (sometimes greatly annoying firefighters), and mate and then lay their eggs in the charred remains. The larvae of the beetles would not be able to cope with the defensive plant compounds in living trees, but they can feed well on burned trees and shrubs because many of the chemical defenses have been destroyed by the fire. The emerging larvae therefore have suitable food available from the burnt vegetation.[9]

6.4 Conclusions

Insects need to be able to ingest those nutrients that their systems cannot synthesize. Plants that phytophagous insects feed on deliver nutrients only in dilute amounts, particularly with respect to sterols. Therefore, insects need to consume large quantities of plant materials. Aside from plants' textures, it is mostly plant chemistry that determines if it is selected by insects as a host. The secondary plant compounds, responsible for taste and smell, and also for color and sometimes toxicity, make insects select or avoid a plant as a food source.

Some laboratory methods can be used to shed light on the nutritional needs and plant preferences of specific insects. Electro-antennograms can measure an insect's response to olfactory stimuli as, *e.g.* triggered by plant natural products. Artificial diet solutions can be used to explore feeding preferences and nutritional needs of insects.

Knowledge about insects' feeding preferences and mechanisms is of interest for the management of invasive insects – but also for attracting desirable insects to a garden.

References

1. L. M. Schoonhoven, J. J. A. van Loop and M. Dicke, *Insect-Plant Biology*, Oxford University Press, Oxford, 2005.
2. D. H. Akey and S. D. Beck, Continuous Rearing of the Pea Aphid, *Acyrthosiphon pisum*, on a Holidic Diet, *Ann. Entomol. Soc. Am.*, 1970, **64**(2), 353.
3. R. F. Chapman, A. E. Douglas and S. J. Simpson, Nutrition, in *The Insects: Structure and Function*, ed. S. J. Simpson and A. E. Douglas, Cambridge University Press, Cambridge, UK, 5th edn, 2013, ch. 4.
4. P. Jolivet, *Interrelationship Between Insects and Plants*, CRC Press, Boca Raton, FL, 1998.
5. E. A. Bernays and R. F. Chapman, *Host-Plant Selection by Phytophagous Insects*, Chapman & Hall, New York, NY, 1994.
6. For example: J. P. Brock and K. Kaufman, *Kaufman Field Guide to Butterflies of North America*, Houghton Mifflin, New York, NY, 2003.
7. P. J. Gullan and P. S. Cranston, *The Insects: An Outline of Entomology*, Wiley-Blackwell, Chichester, West Sussex, UK, 2010.
8. D. L. Dreyer, J. C. Reese and K. C. Jones, Aphid Feeding Deterrents in Sorghum, *J. Chem. Ecol.*, 1981, 7(2), 273.
9. H. Schmitz and H. Bleckmann, The photomechanic infrared receptor for the detection of forest fires in the buprestid beetle *Melanophila acuminata* (Coleoptera: Buprestidae), *J. Comp. Physiol.*, A, 1998, **182**, 647.

7 Plant Galls: Protection and Food for the Young

7.1 Introduction

Some types of insect–plant interactions cause the formation of abnormal growths on plants. When certain insects sting leaves, twigs, trunks, roots, fruits, or flower buds and proceed to lay an egg (or several eggs) into the plant part, a local plant tumor, called a 'gall', is formed. The development of the gall begins when a larva hatches from the egg and starts eating plant material. The gall itself consists of plant cells that grow abnormally, as a consequence of either the mechanical or chemical feeding stimuli by a specific insect.[1] The study of plant galls is known as *cecidology.*

Galls are mostly benign plant tumors and generally do not cause the death of the plant. They clearly benefit the growing insects because they protect the enclosed eggs and the hatching larvae and provide food for the developing insects. As a potential benefit for the host plants, galls encapsulate larval insects that otherwise might be more damaging by openly feeding on plants.[2]

It is common that several insects use a gall, in complex interactions. Often an insect other than the gall-inducer makes its way into the gall and consumes the original larva, and then emerges from the gall. Small wasps called cynipid wasps (Cynipidae) are a major family of gall-making insects. Figure 7.1(a) shows a large gall 'apple' attached to a twig of a valley oak (*Quercus lobata*); the growth had been induced by a California gall wasp (*Andricus quercus-californicus*). Figure 7.1(b) shows very differently shaped galls on the underside of

The Chemistry of Plants and Insects: Plants, Bugs, and Molecules
By Margareta Séquin
© Margareta Séquin 2017
Published by the Royal Society of Chemistry, www.rsc.org

(a)   (b)   (c)

Figure 7.1   Plant galls. (a) Gall 'apple' on valley oak (*Quercus lobata*), induced by a California gall wasp (*Andricus quercus-californicus*). (b) Colorful galls on the underside of a leaf of a blue oak (*Quercus douglasii*), caused by a woollybear gall wasp (*Atrusca trimaculosa*). (c) A manzanita leaf gall aphid (*Tamalia coweni*) caused this gall at the margin of a manzanita leaf (*Arctostaphylos* sp.).

a leaf of a blue oak (*Quercus douglasii*) caused by another type of gall wasp, namely a woollybear gall wasp (*Atrusca trimaculosa*). Some species of aphids can induce gall formations as well. For example, a type of aphid (*Tamalia coweni*, Aphididae) caused the reddish gall along the margin of a manzanita leaf (*Arctostaphylos* sp.) shown in Figure 7.1(c). Other families of gall-inducing insects include sawflies (small wasp-like insects), tiny gall midges (Cecidomyiidae), and thrips (Thysanoptera). A few types of beetles and moths can also cause the growth of galls.[3]

Plant galls exhibit a great diversity of shapes, sizes, and colors (see Figure 7.1). Each species of gall-making insect causes a different type and shape of gall, and on a specific plant. Thus it is possible to identify the galling organism by the shape of a gall. Oak trees (*Quercus* sp.), willows (*Salix* sp.), and plants of the rose family (Rosaceae) are common host plants of galls. But the growths can develop on almost any type of plant. Galls are found on plants worldwide. For example, more than two hundred types of galls are known in Europe alone.[4] Interestingly, the gall-making relationship between specific insects and plant hosts has evolved several times independently, in unrelated insect and plant families. Gall formation is an ancient relationship between insects and plants. Fossilized galls have been found on fossils of tree ferns as far back as the late Carboniferous Period, more than 300 million years ago.[5]

A few gall-inducing insects cause great damage to their host plants. A notorious example is the grape louse or grape phylloxera (*Daktulosphaira vitifoliae*, also known as *Phylloxera vitifoliae*). This insect is related to aphids. It originated in Eastern North America, from vines (*Vitis* sp.) native there that are resistant to the gall-making. Phylloxera has created widespread damage to nonresistant vines in vineyards worldwide, notably to the vineyards of France. During complex life cycles of numerous generations, some grape phylloxera induce galls on grape leaves, whereas others migrate to the roots of grape vines leading to large root swellings and the subsequent death of the vines. In parts of the world where rice is grown, like Southern and Southeast Asia, the rice gall midge (*Orseolia oryzae*) induces pale tubular galls on the shoot terminals, inflicting great damage to rice crops.

While insects are the major gall-inducing organisms, growth of plant galls can also be caused by tiny spiderlike mites, by fungi, or by viruses. Furthermore, bacteria can initiate development of tumors on plants. An example is the widespread and highly damaging *Agrobacterium tumefaciens* that causes crown gall.

People have been intrigued by the curiously shaped, sometimes colorful growths on plants since ancient times. Some galls were used as dyes, for making inks, for medicinal purposes, or even as edible fruits as in the case of figs (*Ficus* sp.).

There are more than eight hundred species of figs (*Ficus* sp.), most of them tropical plants. Different species of *Ficus* have different pollination mechanisms, but all require a specific species of wasp (Family: Agaonidae) for pollination, in complex interactions between fig plants and the insects. The fig wasps are examples of mono-phagous insects, as each species of wasp visits a particular species of *Ficus*.[6] A wild fig plant with fruits is shown in Figure 7.2(a). Volatile attractants produced by the plant attract the specific wasps. The new buds of fig plants (Figure 7.2(b)) house masses of small flowers called

(a)   (b)   (c)

Figure 7.2   Figs. (a) Wild figs on tree (*Ficus* sp.). (b) Young bud of fig. (c) Ripe edible figs, when cut open, show the syconium (masses of former flowers).

a 'syconium'. Tiny female wasps, carrying pollen from another fig plant, enter the emerging buds. During their visit they pollinate some of the fig blossoms and lay eggs. Formation of the fig starts with the developing larvae. Female wasps still in the syconium mate with emerging male wasps, then leave the maturing figs. (Cultivated figs are mostly self-fertilizing.) Figure 7.2(c) shows mature edible figs; when cut open, the masses of former tiny flowers can be seen.

7.2 Galls and Their Chemistry

Galls, especially oak galls, have a high content of tannins, including gallic acid 7.1 and more complex tannins 7.2 (Figure 7.3. See also Chapter 4.6). Tannins from galls have been the sources for brown to black dyes since Roman times and were in use well into the twentieth century. Black inks were produced from oak galls by grinding them up into powders and boiling them in water. This process hydrolyzes the tannins mostly into gallic acid. When an aqueous solution of ferrous sulfate ($FeSO_4$) is added, water-soluble ferrous (Fe^{2+}) gallates are formed that can penetrate into a paper surface. Upon exposure to air, oxygen oxidizes the ferrous ions to ferric (Fe^{3+}) ions. The higher

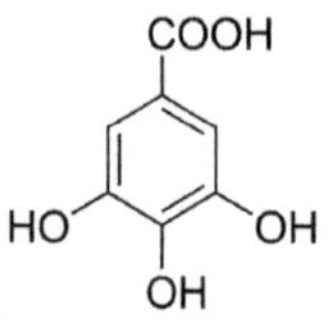

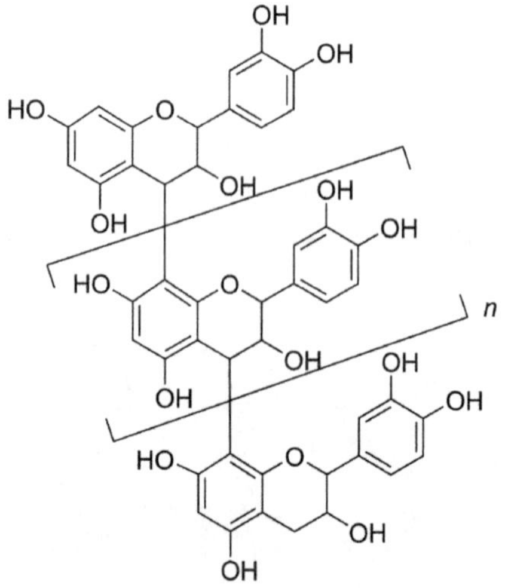

Figure 7.3   Tannins. Simple tannins like gallic acid **7.1** and more complex tannins as shown in **7.2** are common phenolics in galls, particularly in oak galls.

charged ions strongly bind to the phenolic groups of gallic acid, and water-insoluble polymeric ferric tannates are formed. The resulting black ink is permanent on paper. The high acidity of this ink, however, has been a problem as it causes the deterioration of old manuscripts.

7.3 Mysteries of Gall Formation

What is the nature of the signal that induces the formation of a plant gall? The exact mechanisms of the induction and growing of galls are still largely unknown, although it has long been observed that the sting of an insect into a plant part, and the subsequent egg-laying, stimulate local abnormal plant growth that forms the gall. Ground-breaking work by Y. Suzuki and coworkers suggests that secretions from the insect, when it stings a plant part and lays an egg into it, trigger the local production of plant hormones or plant growth factors, like auxins and cytokinins.[7]

Figure 7.4 shows the structures of important plant hormones that generally stimulate and regulate plant growth. (Earlier we encountered ethylene, $CH_2=CH_2$ **2.33**, as a gaseous ripening hormone.) Indoleacetic acid, specifically indole-3-acetic acid (abbreviated as IAA) **7.3**, is present in all parts of a plant, particularly those that are involved in active growth. Its structure closely resembles the amino acid tryptophan **7.4**, from which plants can synthesize the plant growth factor. Indoleacetic acid belongs to a group of plant hormones that actively promote growth of plant cells, called *auxins*. It is the most abundant and most important naturally occurring auxin. Cytokinins are plant growth factors that regulate cell division in plants, primarily in actively-dividing plant tissues, like root tips, fruits, or leaves.[8]

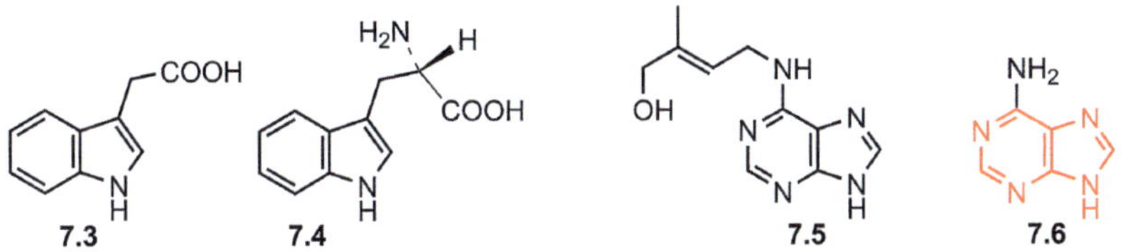

7.3   7.4   7.5   7.6

Figure 7.4   Plant hormones or growth factors. The plant hormone indole-acetic acid (IAA) **7.3** is the principal auxin of higher plants. Its structure is closely related to the amino acid tryptophan **7.4**. Zeatin **7.5** is a common cytokinin. Its structure is related to adenine **7.6**. (The purine system is highlighted in red.)

They act in concert with auxins to regulate cell differentiation. Zeatin **7.5** is a widespread cytokinin that was first isolated from corn or maize (*Zea mays*). Note that its chemical structure is related to adenine **7.6**, one of the purine bases in DNA; the purine structure is highlighted.

The studies by Y. Suzuki and coworkers involved a sawfly species (*Pontania* sp.) that induces leaf galls on its willow host plant (*Salix japonica*). At the time of egg-laying into willow leaves, the adult sawflies inject the contents of their glands which were found to contain a precursor of a cytokinin. The development of the gall is maintained by the growing larvae through the production of high concentrations of indoleacetic acid **7.3** and the cytokinin zeatin **7.5**. The researchers discovered that the emerging larvae were able to synthesize indole-acetic acid from tryptophan **7.4** themselves. The findings of this research suggest that the gall-inducing insects have evolved mechanisms that make use of plant growth mechanisms in order to produce galls. Studies on the details of these processes continue, as well as investigations on how widespread similar mechanisms are in gall formations.

Many further questions remain, *e.g.* with regards to the formation of the highly diverse shapes and colorations of galls, induced by specific insects on specific plants. Can insects genetically reprogram the growth of a plant part? Is the local genetic information changed? Such a mechanism is known from the crown gall bacterium that has been found to transfer its own small piece of deoxyribonucleic acid (DNA) into the plant cell and this becomes part of the plant's DNA. The altered DNA codes for genes that produce auxin and cytokinin, inducing the growth of the crown gall tumors.[9]

7.4 Conclusions

Galls are the result of highly specific and complex interactions of insects with plants. People have long been fascinated by these plant tumors. Their shapes, their attending insects and host plants have been studied extensively. But the actual causes of gall formation, their chemistry and biochemistry, are still largely unknown.

Advances in molecular biology and new techniques in instrumental analysis should make it possible to answer many questions regarding gall formation in the not too distant future. Aside from the fascination with galls, an understanding of the causes of the gall-making processes is of special interest in agriculture with respect to galls harmful to crops.

References

1. A. Raman, Insect-Plant Interactions: The Gall Factor, in *All Flesh is Grass – Plant-Animal Interrelationships*, ed. J. Seckbach and Z. Dubinsky, Springer, Dordrecht, 2011, pp. 121–146.
2. P. Jolivet, *Interrelationship Between Insects and Plants*, CRC Press, Boca Raton, 1998.
3. R. Russo, *Field Guide to Plant Galls of California and other Western States,* California Natural History Guides, UC Press, Berkeley, 2006.
4. P. Bryant, Inducers, parasitoids, and inquilines: Life inside the plant gall, *Fremontia*, 2013, **41**(3), 14.
5. C. C. Labandeira, Early history of arthropod and vascular plant associations, *Annu. Rev. Earth Planet. Sci.*, 1998, **26**, 329.
6. P. J. Gullan and P. S. Cranston, *The Insects: An Outline of Entomology*, Wiley-Blackwell, Chichester, West Sussex, UK, 2010.
7. H. Yamaguchi, H. Tanaka, M. Hasegawa, M. Tokuda, T. Asami and Y. Suzuki, Phytohormones and willow gall induction by a gall-inducing sawfly, *New Phytol.*, 2012, **196**(2), 586.
8. R. F. Evert and S. E. Eichhorn, *Raven Biology of Plants*, W. H. Freeman, New York, NY, 8th edn, 2012.
9. A. Costacurta and J. Vanderleyden, Synthesis of phytohormones by plant-associated bacteria, *Crit. Rev. Microbiol.*, 1995, **21**(1), 1.

8 Insects That Use Plant Defenses for Their Own Protection

8.1 Introduction

Diverse chemical defenses in plants, with strong odors, bitter tastes, or toxic properties, deter many insects from eating the plants (see Chapter 4). Yet, there are numerous examples where insects have not only adapted to feed on such plants, in spite of the defensive compounds, but they have also developed mechanisms to store the ingested plant defenses in their bodies and use them for their own protection. Predators like birds that eat the insects become sick and learn to avoid them. Bright warning colors (*i.e.* the aposematic colors) frequently alert potential predators about the toxicity or poor taste of these insects. Examples of insects that store toxins and exhibit warning colors have appeared earlier in this book in various contexts, *e.g.* in Figure 1.3 in Chapter 1 and Figure 5.10 in Chapter 5. You will find more examples in this chapter.

A variety of mechanisms enable certain insects to store plant toxins. Some insects sequester the plant defenses in select, well-separated parts of their bodies. Other insects can reduce or eliminate the toxic properties of the compounds by modifying them to related, non-poisonous compounds. The defensive compounds may be converted back to the toxins in animals preying on the insects. Many mono-phagous and oligophagous insects, *i.e.* insects that specifically feed on one type of plant or a few related plants, are connected with host plants that feature distinct chemical defenses. The numerous sequences of short life cycles strongly support natural selection of

The Chemistry of Plants and Insects: Plants, Bugs, and Molecules
By Margareta Séquin
© Margareta Séquin 2017
Published by the Royal Society of Chemistry, www.rsc.org

those insects that can adapt to otherwise deterring plant compounds and that can even use them in their own defense.

Many types of butterflies and moths are known to acquire toxicity from plants because their larvae, the caterpillars, feed on poisonous plants. The emerging adults mostly sip flower nectars, but they still contain enough of the chemical defenses in their bodies to make predators that eat them ill. Aphids, members of the superfamily Aphidoidea, are widespread sap-sucking insects, and many of them are plant pests in agriculture.[1] Their numerous life cycles per year, especially in mild climates, allow them to quickly adjust not only to environmental conditions, but also to host plants and their chemistry. Many plants that contain toxins have specialized aphids that feed on them in spite of the defensive compounds. In the process these aphids ingest chemical protection for themselves.

This chapter presents classic examples of *coevolution* between insects and plants. Through evolution many plants developed chemical defenses as a protection against herbivorous insects. In response, many phytophagous insects have counteradapted to the plant defenses and sometimes even use them for their own defense. Compare these mutual adaptations with those that evolved – and that keep evolving – between plants and their insect pollinators (Chapter 2).

The following chapter sections describe specific examples of insects that obtain chemical protection from plants. The descriptions also point out the chemical characteristics of the defensive compounds involved.

8.2 Monarchs, Milkweeds, and Cardiac Glycosides

Monarch butterflies (*Danaus plexippus*) have long fascinated people because of their far-reaching seasonal migrations, but also because of their toxicity to birds that try to eat them. The connection between milkweeds (*Asclepias* sp.) serving as food plants for the monarch caterpillars and the toxicity of the adult butterflies was suggested by E. B. Poulton as early as 1914.[2] This work presents a famous case of the intersection of biology and chemistry and illustrates a classic example of insects specializing on toxic host plants and using the poisons for their own protection. Milkweed plants (*Asclepias* sp., Family Apocynaceae) are native to North and Central America. Their common name alludes to the white latex in all parts of the plants (compare Chapter 4.8). Their systematic name, *Asclepias*, was assigned by the botanist Carl Linnaeus. It is related to Asclepius, the

Greek god of healing, because of the heart-active properties of milk-weed plants. Ongoing studies on how insects are able to detoxify and thus tolerate the chemical defenses demonstrate continued interest in these insect–plant interactions.[3]

Figure 8.1 shows various stages of the monarch's life cycle: a caterpillar eating milkweed leaves (Figure 8.1(a)), the fully developed butterfly feeding on flower nectar (Figure 8.1(b)), and a cluster of overwintering monarch butterflies (Figure 8.1(c)), a phenomenon that is part of the annual migration of the butterflies. Note the warning colors of the insects. The mechanisms that enable monarch butterflies and their caterpillars to store the toxic defenses are still under investigation.

Aside from the monarch butterflies there is a suite of other insects that also feed on *Asclepias* species. Milkweed beetles (*Tetraopes tetrophthalmus*, Figure 1.3(b)), milkweed bugs (*Oncopeltus fasciatus*, Figure 1.6(c)), and milkweed (or oleander) aphids (*Aphis nerii*, Figure 4.17(b)) all ingest the plants' toxins without harm and use them for their own defense. Note that all these insects display bright warning colors. Leaves and stems of milkweeds contain a particularly high concentration of the toxins in their "milk" or latex, with the highest pressure of the latex in the veins of the leaves (compare Chapter 4.8). These insects have adapted to ingest small portions of the leaves, often cutting plant veins first to release the pressure. Thus the dose of the ingested toxins is low and tolerable for the insects. As a further benefit, the food plants survive.

The toxicity of the plants' latex is mainly due to cardiac glycosides. As the name 'cardiac' implies, these compounds affect (and inhibit) the proper functioning of the heart muscle. The determination of the

(a)   (b)   (c)

Figure 8.1 Life stages of the monarch butterfly (*Danaus plexippus*). (a) Monarch caterpillar feeding on milkweed (*Asclepias* sp.). (b) Adult monarch butterfly feeding on nectar of *Zinnia* flowers. (c) Cluster of overwintering monarch butterflies.

structures of these compounds – with thoughts of potential medicinal applications – was the topic of intense research in the 1960's.[4] The chemical structures of cardiac glycosides consist of three general components: a steroid backbone structure (shown with conventional numbering and labeling in **8.1** in Figure 8.2), a sugar moiety attached to C3 of the steroid ring system, and a characteristic ring attached to C17. Check each of the structures shown in Figure 8.2 for these three components. Incidentally, the name 'steroid' is derived from 'steros' (Greek) which means rigid. Steroid molecules are rigid ring systems, with their rings connected in defined steric arrangements.

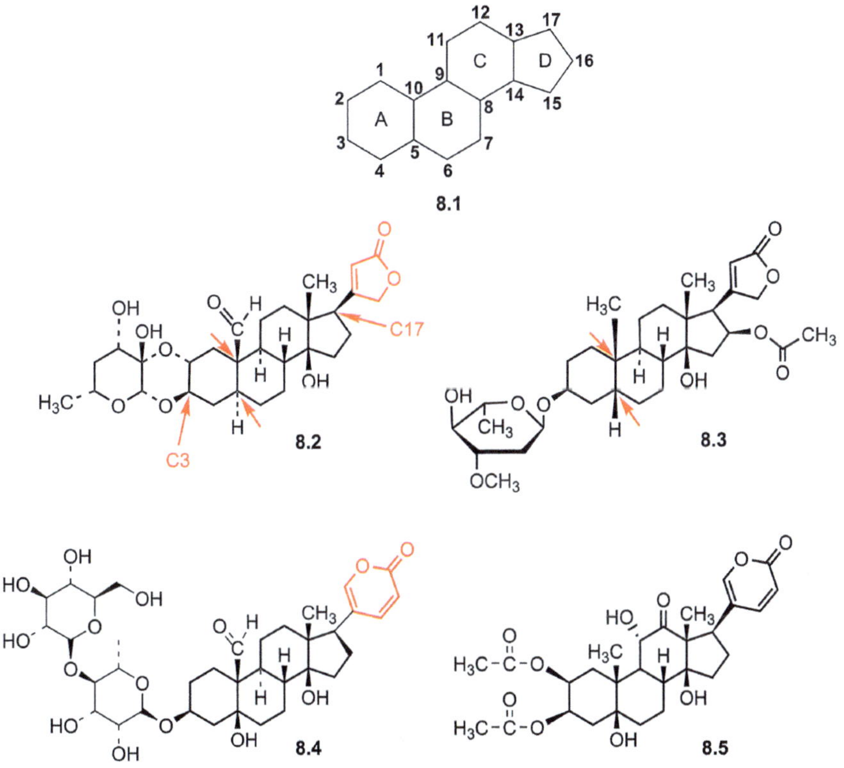

Figure 8.2 Cardiac glycosides. Structure **8.1** shows the characteristic ring system of steroids, with conventional numbering and labeling. Calotropin **8.2** from milkweeds, and oleandrin **8.3** in oleanders are toxic cardenolides. Hellebrin **8.4**, from the plant *Helleborus niger*, and lucibufagin **8.5**, from fireflies, are both bufadieno-lides. The characteristic rings attached to C17 in cardenolides and bufadienolides are highlighted in **8.2** and **8.4**, respectively.

There are two groups of cardiac glycosides: cardenolides and bufadienolides. They differ in the type of ring that is attached to C17. The toxic compounds in milkweeds belong to the group of steroids called *cardenolides*.[5] A characteristic five-membered ring, called a butenolide, is attached to C17 in their molecules. This ring is highlighted in the molecular structure of calotropin **8.2** (Figure 8.2), the major toxin in milkweeds. Note that rings A and B in calotropin are connected in *trans* fashion (pointed out by the arrows in **8.2**). This is typical of milkweed cardenolides. Recall that 'glycoside' generally means that molecules have a carbohydrate or sugar moiety, with several polar functional groups like OH groups attached; they make the entire molecule more soluble in polar media like water. The non-sugar part of the glycosides, the *aglycone*, is the part that is responsible for the inhibiting action on the heart muscle in cardiac glycosides. Nevertheless, the sugar moiety, due to its contribution to polarity of the glycoside, affects the uptake of the cardenolide in the gut and with this its toxicity as well. Oleanders (*Nerium oleander*, Apocynaceae family, Figure 8.3(a)) contain the toxic cardenolide oleandrin **8.3**. Note that rings A and B in the steroid section of oleandrin have a *cis* connection, as pointed out by arrows in Figure 8.2. Another well-known plant source of cardenolides is foxglove (*Digitalis* sp., Figure 8.3(b)). Both oleanders and foxglove plants have their specific aphids that feed on them.

Bufadienolides are another group of cardiac glycosides. They feature a characteristic six-membered lactone ring (highlighted in **8.4**).

(a)   **(b)**   **(c)**

Figure 8.3 Examples of plants with cardiac glycosides that attract insect specialists. (a) Oleander (*Nerium oleander*). (b) Foxglove (*Digitalis* sp.). (c) Christmas rose (*Helleborus niger*) with hellebore aphids (*Macrosiphon hellebori*).

Hellebrin **8.4**, from *Helleborus* plants, like the Christmas rose (*Helleborus niger*), is a toxic, cardiotonic bufadienolide. Yet, a type of aphid (*Macrosiphum hellebori)* feeds on the plants unimpeded (Figure 8.3(c)). Figure 8.2 also shows the structure of lucibufagin **8.5**, a bufadienolide that acts as a chemical defense in fireflies. This compound was shown earlier in connection with chemical defenses of insects in Chapter 5.5.

Cardenolides and bufadienolides are found in plants and animals. While some insects ingest them directly from plant sources, several insect species can synthesize them from steroids in their own bodies. (The latter also applies to other animals, like toads, *Bufo* sp.)[6]

The elucidations and subsequent syntheses of the complex structures of cardiac glycosides, including the numerous defined chiral centers, provided major challenges to researchers. Plant cardenolides, like the heart glycosides from *Digitalis*, and compounds derived from them, have important applications in human medicine and are used to treat various heart conditions.

8.3   *Heliconius* Butterflies, Passion Vines, and Cyanides

Butterflies of the genus *Heliconius* are colorful tropical insects in the large family of Nymphalidae or brush-footed butterflies. Their host plants are cyanogenic passion vines (*Passiflora* sp.), a mostly tropical plant genus with complex flowers and diverse vegetative parts. The connection between *Heliconius* butterflies and their toxic food plants has been well-researched.[7] It is another example of coevolution between insects and toxic plants in which the respective larvae (the caterpillars) and the emerging adult butterflies have adapted to tolerate the chemical plant defenses and to store them for their own defense. Figure 8.4(a) shows a postman butterfly (*Heliconius melpomene*) that is found in Central America. Its caterpillar exclusively feeds on the leaves of passion vines, like those of the red passion vine shown in Figure 8.4(b). These caterpillars have soft spines and are poisonous to birds.[8] A related butterfly, within the same family, is the Gulf fritillary (*Agraulis vanillae*) shown in Figure 5.7(b). Its larvae also feed on passion vines. The caterpillars are bright orange in color and covered in rows of black spines; they are poisonous to birds. The cultivation of passion vines has enabled the Gulf fritillary to extend its range northwards into California.[9]

Figure 8.4   Passion vines and butterflies. (a) One of the *Heliconius* or postman butterflies (*Heliconius melpomene rosina*) from Central America. Its caterpillars feed on leaves of passion vines (*Passiflora* sp.). (b) Red passion vine (*Passiflora* sp.) flower and leaves.

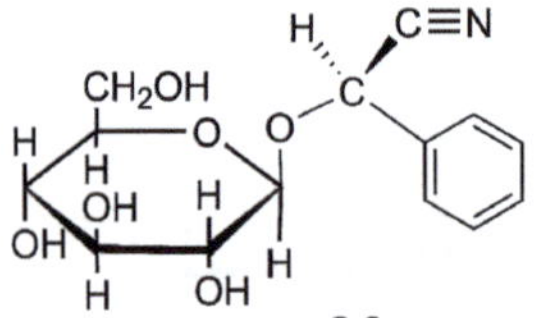

Figure 8.5   Examples of cyanogenic glycosides found in passion vines (*Passiflora* sp.). Prunasin **8.6** is a very common cyanogenic glycoside found in many different plants. Passibiflorin **8.7**, specific to *Passiflora*, features an unusual cyclopentene ring (highlighted) and glycoside pattern.

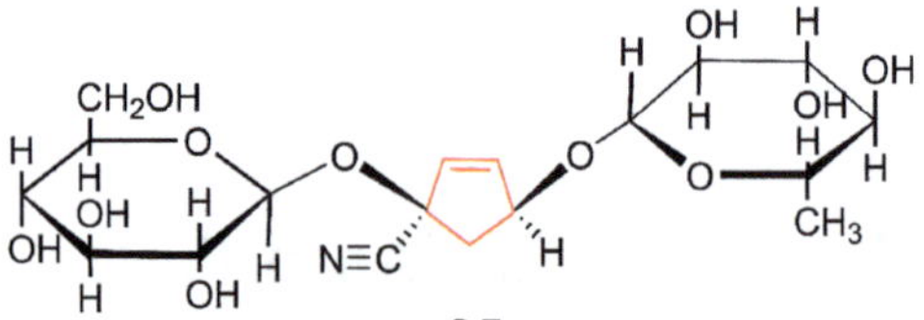

Generally, these butterflies lay their eggs on green passion vine leaves, thus securing a food source for the emerging caterpillars. As an interesting tidbit of information on evolutionary paths, some types of passion vines feature small, colored nubs on their leaves which resemble the butterflies' eggs. This seems to prevent the butterflies from laying too many eggs on any single plant.

While feeding on the leaves of passion vines, the caterpillars ingest cyanogenic glycosides from the plants (compare Chapter 4.4). Some examples of cyanogenics that occur in passion vines are shown in Figure 8.5. Aside from common compounds, like prunasin **8.6**, that occur in many other types of plants as well, several unusual cyanogenics are specifically found in *Passiflora*. They feature five-membered cyclopentene rings (highlighted in **8.7**), with sometimes extended carbohydrate moieties. An example of such a compound is passibiflorin **8.7**.

As was addressed in Chapter 4.4, cyanogenic glycosides from plants are generally hydrolyzed by a β-glucosidase from the plant if an insect starts feeding on it. The reaction produces a spontaneous release of hydrogen cyanide (Figure 4.9). So, how can *Heliconius* caterpillars ingest the toxins without harm and sequester them for their own defense? It has been found that these insects have enzyme systems that can inactivate the plant enzymes, thus allowing the uptake and storage of the un-hydrolyzed cyanogenic glycoside. Aside from ingesting the cyano-genics directly from the host plants, *Heliconius* butterflies can also synthesize them *de novo* from amino acid precursors. β-Glucosidase enzymes are stored in separate body parts of the insects. When *Heliconius* caterpillars or butterflies are attacked (*i.e.* bitten into), the enzymes come into contact with the cyanogenics, with the result of the release of hydrogen cyanide.[10,11]

8.4   Pipevine Butterflies, Pipevines, and Alkaloids

Another well-studied example of an insect that ingests toxins from specific plants and stores them as chemical defense in its own body is the pipevine swallowtail (*Battus philenor*), a butterfly native to North and Central America. Its caterpillar (Figure 8.6(a)), with a distinct black body and decked with orange spines, feeds exclusively on leaves of toxic pipevines (*Aristolochia* sp.), while the adult butterfly, still containing the plant toxins, feeds on flower nectars (Figure 8.6(b)). Nevertheless, the adult butterfly remains closely connected with

(a)   (b)

Figure 8.6   Pipevine butterfly (*Battus philenor*), larva and adult, and host plants. (a) Black and orange caterpillar of *Battus philenor*, feeding on leaves of California pipevine (*Aristolochia californica*). (b) Adult pipevine swallowtail butterfly feeding on nectar of blue dick brodiaea flowers (*Dichelostemma capitatum*).

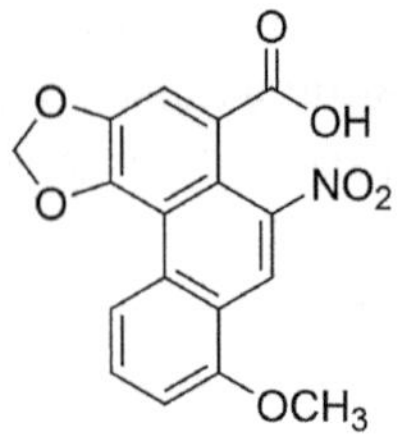

8.6

Figure 8.7   Aristolochic acid **8.6**, a weakly acidic alkaloid found in pipevines (*Aristolochia* sp.).

pipevine plants, laying its eggs on leaves of *Aristolochia* sp. The emerging caterpillars then start feeding on the leaves that contain bitter alkaloids. While the caterpillars tolerate the toxins, they do not particularly like the bitter taste, especially of mature pipevine leaves where the alkaloid content is considerably higher than in young leaves. Therefore, the caterpillars tend to ingest only small parts of the leaves, thus allowing the plants to survive – and to continue to serve as host plants.

The main alkaloid in pipevines is aristolochic acid **8.6** (Figure 8.7). It belongs to a group of carcinogenic, mutagenic, and nephrotoxic compounds, the aristolochic acids, that differ by slight variations of the functional groups attached. Aristolochic acid is an unusual type of alkaloid. While its molecular structure contains the required nitrogen, the nitrogen atom is part of a nitro group (NO_2), and not part of a ring (which is the usual arrangement). Aristolochic acid is a weak acid as its name implies. (Note its carboxylic acid group.) This is unlike most alkaloids that characteristically have alkaline properties due to basic nitrogen atoms in their structures. (Compare Chapter 4.7.)

8.5   Moths, Ragworts, and *Senecio* Alkaloids

Groundsels (*Senecio vulgaris*, Figure 8.8(a)) and ragworts (*Senecio jacobaea*) are common, weedy plants in the Compositae family (Asteraceae). They are characteristic of disturbed places. The plants are native to northern Eurasia and have been introduced to North America, New Zealand, and Australia where they are considered invasive. Prolific seed production as well as effective chemical defenses, in the form of specific alkaloids, contribute to the success of these plants. The general name of *Senecio* alkaloids describes their common occurrence in plants of the genus *Senecio*. These alkaloids are also

(a) (b) (c)

Figure 8.8   *Senecio* plants and specialized moths. (a) Groundsels (*Senecio vulgaris*) are weedy plants growing in disturbed places. (b) Brightly colored caterpillar of the cinnabar moth (*Tyria jacobaeae*) feeding on ragwort (*Senecio jacobaea*). (Photo by Quartl. Wikimedia Commons. https://upload.wikimedia.org/wikipedia/commons/e/e3/Tyria_jacobaeae_qtl1.jpg, (accessed October 2016).) (c) Adult cinnabar moth displaying aposematic colors. (Photo by Wikimedia Commons. https://upload.wikimedia.org/wikipedia/commons/a/ae/Tyria_jacobaeae-04_%28xndr%29.jpg, (accessed October 2016).)

typically found in plants of the borage family (Boraginaceae), like sweet comfrey (*Symphytum officinale*). Another name for them is pyrrolizidine alkaloids, referring to the characteristic ring systems in their chemical structures. Pyrrolizidine or *Senecio* alkaloids are hepatotoxic, *i.e.* damage the liver. They can pose a danger to livestock and horses if ingested in high enough doses, as in hay. The inflorescences of *Senecio* plants offer important food to many pollinators. Honey from bees visiting flowers of *Senecio* plants contains small amounts of the alkaloids.

Animals in general, and insects in particular, avoid eating leaves and stems of fresh *Senecio* plants because of their bitter taste due to the alkaloids. But caterpillars of the cinnabar moth (*Tyria jacobaeae*) have adapted to them as host plants and at times infest them. They can slightly modify the toxins and sequester them, and thus obtain their own chemical defense. Note the aposematic colors of both the larvae (Figure 8.8(b)) and the adult, a day-flying moth (Figure 8.8 (c)). The striking colors alert potential predators. Furthermore, the alkaloids ingested from *Senecio* plants are passed by the adult male moths to the females during mating. The females then transmit the toxins to their eggs, which become distasteful to predators. Cinnabar moths have been introduced to North America and New Zealand as biological control of invasive *Senecio* plants.[12,13]

The name 'pyrrolizidine alkaloids' refers to the typical bicyclic ring system that is part of their structures (highlighted in **8.7**, Figure 8.9), with a nitrogen atom as part of the rings. In plants, *Senecio* alkaloids occur as ring-forming diesters. (Note the ester functional groups, –COOC–.) The diesters are large-ring or 'macrocyclic' esters, formed from 'necine bases', with the pyrrolizidine ring system, and two carboxylic acid groups. Senecionine **8.7** is an example of a common pyrrolizidine alkaloid in *Senecio* plants. Its toxicity is described as having an LD_{50} i.v. in mice: 64.12 ± 2.24 mg kg^{-1}. (In comparison, aspirin is listed with an LD_{50} orally in mice: 1.5 g kg^{-1}.)[14] When non-adapted animals, like many mammals and most insects, ingest alkaloids like senecionine, the esterified pyrrolizidine alkaloids are directly hydrolyzed in their metabolisms. These reactions lead to poisonous pyrrolizidines, like retronecine **8.8** and, furthermore, to even more toxic, oxidized forms **8.9** with a pyrrol ring. These lipophilic alkaloids passively diffuse through biological membranes. All are hepatotoxic.[15,16] Adapted insects, like the cinnabar moth, can oxidize the toxic pyrrolizidines to the non-toxic *N*-oxides **8.10**. The latter are hydrophilic, salt-like compounds (note the charges) that cannot diffuse through membranes without a specific membrane carrier. Adapted insects do have these specific carriers which make

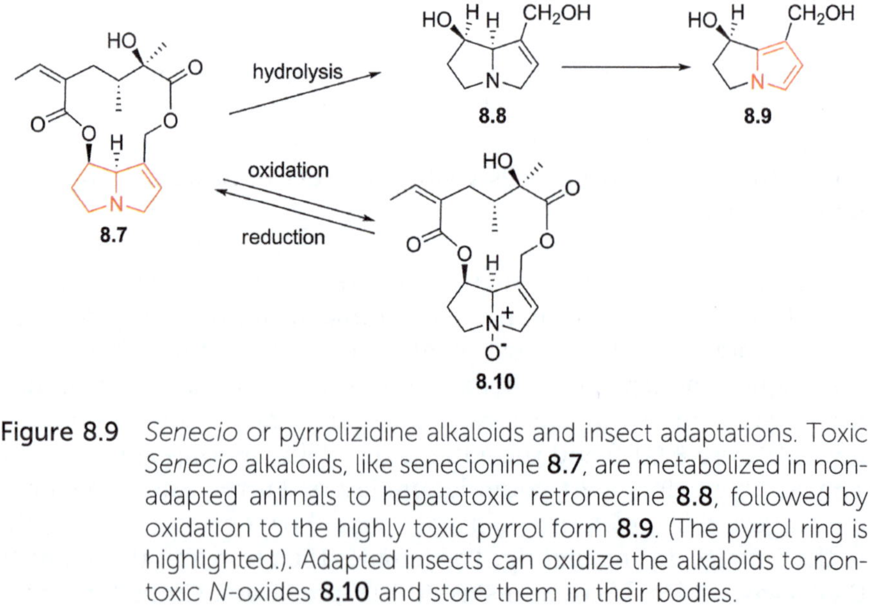

Figure 8.9 *Senecio* or pyrrolizidine alkaloids and insect adaptations. Toxic *Senecio* alkaloids, like senecionine **8.7**, are metabolized in non-adapted animals to hepatotoxic retronecine **8.8**, followed by oxidation to the highly toxic pyrrol form **8.9**. (The pyrrol ring is highlighted.). Adapted insects can oxidize the alkaloids to non-toxic *N*-oxides **8.10** and store them in their bodies.

the transport of N-oxides across membranes possible. By these mechanisms adapted insects avoid self-poisoning and can store the compounds as potential chemical defenses in their bodies.[17]

8.6 Conclusions

This chapter presented examples of insects that, through evolution, have adapted to chemical plant deterrents and are able to use them for their own defense. Insects that specialize on toxic host plants are likely suspects to feature such adaptations, especially when displaying striking colors. This chapter also pointed to the diversity of defensive plant secondary metabolites and addressed some mechanisms that allow insects not only to cope with the toxins, but also to store them in their systems. Many details of the processes that allow insects to ingest and accumulate the toxins for their own defense have not yet been explained and are topics of ongoing investigations.

References

1. D. Iluz, The Plant-Aphid Universe, in *All Flesh is Grass – Plant-Animal Interrelationships*, ed. J. Seckbach and Z. Dubinsky, *Cellular Origin, Life in Extreme Habitats and Astrobiology*, Springer, Dordrecht, 2011, vol. 16, pp. 93–118.
2. S. B. Malcolm, Milkweeds, monarch butterflies and the ecological significance of cardenolides, *Chemoecology*, 1994/1995, **5/6**, 101.
3. S. Dobler, G. Petschenka and H. Pankoke, Coping with toxic plant compounds–the insect's perspective on iridoid glycosides and cardenolides, *Phytochemistry*, 2011, **72**, 1593.
4. T. Reichstein, J. Voneuw, J. A. Parsons and M. Rothschild, Heart poisons in the monarch butterfly, *Science*, 1968, **161**, 861.
5. A. A. Agrawal, G. Petschenka, R. A. Bingham, M. G. Weber and S. Rasmann, Toxic cardenolides: chemical ecology and coevolution of specialized plant-herbivore interactions, *New Phytol.*, 2012, **194**, 28.
6. L. Krenn and B. Kopp, 'Bufadienolides from animal and plant sources', *Phytochemistry*, 1998, **48**(1), 1.
7. K. C. Spencer, Chemical mediation of coevolution in the *Passiflora – Heliconius* interaction, in *Chemical Mediation of Coevolution*, ed. K. Spencer, Academic Press, London, 1988, p. 167.

8. C. L. Henderson, *Butterflies, Moths, and Other Invertebrates of Costa Rica*, University of Texas Press, Austin, TX, 2010.

9. J. P. Brock and K. Kaufman, *Field Guide to Butterflies of North America*, Houghton Mifflin, New York, NY, 2003.

10. H. S. Engler-Chaouat and L. E. Gilbert, '*De novo* Synthesis vs. Sequestration: Negatively Correlated Metabolic Traits and the Evolution of Host Plant Specialization in Cyanogenic Butterflies', *J. Chem. Ecol.*, 2007, **33**, 25.

11. M. M. Hay-Roe and J. Nation, Spectrum of Cyanide Toxicity and Allocation in *Heliconius erato* and *Passiflora* Host Plants, *J. Chem. Ecol.*, 2007, **33**, 319.

12. P. J. Gullan, P. S. Cranston, *The Insects: An Outline of Entomology*, Wiley-Blackwell, Chichester, West Sussex, UK, 2010.

13. J. B. Harborne, *Introduction to Ecological Biochemistry*, Academic Press, London, 4th edn, 1993.

14. *The Merck Index: An Encyclopedia of Chemicals, Drugs, and Biologicals*, ed. M. J. O'Neil, P. E. Heckelman, C. B. Koch and K. J. Roman, Merck & Co., Inc., Whitehouse Station, NJ, 14th edn, 2006.

15. T. Hartmann, Plant-derived secondary metabolites as defensive chemicals in herbivorous insects: a case study in chemical ecology, *Planta*, 2004, **219**, 1.

16. T. Hartmann, Chemical ecology of pyrrolizidine alkaloids, *Planta*, 1999, **207**, 483.

17. A. R. Mattocks, *Chemistry and Toxicology of Pyrrolizidine Alkaloids*, Academic Press, London, 1986.

9 Insects That Provide Protection for Plants

9.1 Introduction

There are many different kinds of mutualistic interactions between plants and insects, *i.e.* mechanisms that mutually benefit plants and insects, and very different services are exchanged in them. In Chapter 2 we addressed the mutualistic relationships between flowering plants and their insect pollinators. Their mechanisms are distinctly different from those that involve insects providing defense for plants. The following examples describe plant features that lure protective insects.

Vast numbers of insects live inside plant structures. Most often, insects create their living spaces or nesting sites by destroying or altering plant tissues, *e.g.* by boring holes into wood or by inducing the formation of galls. In contrast, some specialized plants independently provide shelters and hiding places, without insects initiating them. Offerings of food for the insects are frequently added lures. In return, the attracted insects protect the plants by fending off herbivorous insects (and other animals) and by pruning weeds that encroach on the host plants. In addition, the insect inhabitants produce nitrogenous wastes which can serve as fertilizers for the plants. As a result, these plants are able to live on nutrient-poor soils, such as those found in tropical forests. Besides benefiting from the fertilizing waste products, plants that host defensive insects need to expend fewer resources to produce defensive compounds in order to ward off

The Chemistry of Plants and Insects: Plants, Bugs, and Molecules
By Margareta Séquin
© Margareta Séquin 2017
Published by the Royal Society of Chemistry, www.rsc.org

herbivores. The associations between plants and insects are mutually beneficial.

Many of the associations are *multipartite*; they involve the interactions of multiple types of organisms, like a plant and several types of insects. Such associations have long been observed, and increasing numbers are being studied in detail. The best-known examples are between ants and their host plants.[1,2]

9.2   Ant Plants

Numerous unrelated plants have beneficial relationships with ants. These plants are known as ant plants, or *myrmecophytes*. They are common in tropical habitats where soils typically have a low content of plant nutrients, with only a thin layer of soil produced from rapidly-decaying organic materials. Plants in these environments require mechanisms that make survival possible. Many tropical rain forest plants, especially trees, shrubs, and vines, have evolved mutualistic relationships with ants that defend the plants from herbivores and strangling vines and that also provide plant fertilizers with their waste products.

Ant plants attract the insects by providing shelter and food. Some ant plants feature small chamber-like structures, called *domatia* (from 'domus' for house), in the form of hollow stems, pockets on leaf surfaces, or hollow thorns, sometimes adorned with hair-tufts. The domatia serve as hiding places for the ants. In addition, myrmecophytes attract ants by providing various sources of food. This may be in the form of special *food bodies* that can be located on the stems or at the base or the tips of leaves and that supply nutrients for the insects. In addition, *extrafloral nectaries* may be present that offer sweet, sugary solutions. Some ant species obtain their food exclusively from their host plant. In return, the ants safeguard the plants.

Ants provide excellent protection from herbivorous animals and suffocating vines. Ants living on an ant-loving or *myrmecophilous* plant aggressively attack any animal that comes into contact with it. They can quickly kill foreign intruding insects and other small invertebrates. They can also be extremely irritating to larger vertebrates, including humans. (Tropical biologists are well-advised to recognize ant plants and avoid any contact with them, because touching the plants can cause a fierce attack by the defending ants.) Mutualistic ants also regularly trim encroaching vines.

Several hundred types of ant plants are known, with examples as diverse as acacia trees, orchids, and ferns, and more examples are being discovered. The tropical areas of the Americas and of Southeast Asia are especially rich in myrmecophytes. Macaranga trees (*Macaranga* sp., Euphorbiaceae) are common pioneer plants in the tropical forest understory from Southeast Asia; they also grow in tropical areas of Australia and Africa. Macaranga trees attract ants with food bodies as well as with sweet secretions. The ants find shelter in the hollow stems of the trees and attack herbivorous insects. Other well-known and well-researched examples of myrmecophytes are acacias (*Acacia* sp., Mimosaceae) from Africa and Central America and trees of the New World genus *Cecropia* (Urticaceae or Cecropiaceae) (Figure 9.1).

Cecropia trees are abundant in tropical rain forests of Central America, with about one hundred different species, most of which are ant plants. Figure 9.1(a) shows the tree canopy of a *Cecropia obtusifolia*. *Cecropia* trees are well-known for their mutualism with *Azteca* ants. The insects live well-protected inside the stems and have their nests there. Aside from the structural features, they lure ants with attractive nutrients in special food bodies, with varying compositions in different plants. In the case of *Cecropia*, food bodies located at the base of the leaf stalks (Figure 9.1(b)) provide glycogen, a carbohydrate. Glycogen **9.1** (Figure 9.2) is the main storage polysaccharide in animal cells and fungi; it is an unusual plant product. Starch, on the other hand, is the most common storage carbohydrate in plants

(a) (b) (c)

Figure 9.1   Ant plants (myrmecophytes). (a) Leaf canopy of a *Cecropia* tree (*Cecropia obtusifolia*). (b) The bases of leaf stalks of *Cecropia* trees feature food bodies that lure ants. (c) Branches of a bullhorn acacia (*Acacia* cornigera), with hollow thorns and light-colored food bodies at the tips of the leaflets.
(Photo by Stan Shebs. https://upload.wikimedia.org/wikipedia/commons/4/4f/Acacia_cornigera_2.jpg, (accessed October 2016)).

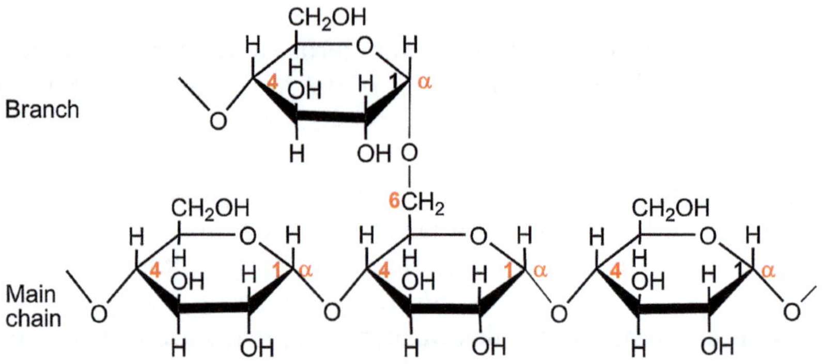

Figure 9.2 Glycogen. A short segment of the polysaccharide glycogen **9.1**, showing the 1,4-α-linked subunits of glucose and a 1,6-α-linked branch.

(Chapter 2.10). Starch build-up takes place in the leaves. One type of *Cecropia* trees, *Cecropia peltata*, produces both storage polysaccharides, starch and glycogen, in a rare example of an organism that can synthesize either polymer.[3] Quantitative measurements have shown that glycogen accumulates to high levels in the specialized food bodies. It provides nutritious, easily-accessible food for the ants.

Glycogen **9.1** (Figure 9.2) has close structural similarities with starch. Recall that starch consists of two polymers, the straight-chained amylose and the branched amylopectin, both polymers with D-glucose as the monomers (Chapter 2.10 and Figure 2.23). Glycogen **9.1** is similar in structure to the branched amylopectin. The D-glucose monomers are connected by 1,4-α-glycosidic linkages. In addition, 1,6-linkages create branching points between main chains and side branches. (The designations of 1,4 and 1,6 designate that carbon atom 1 in one glucose ring is connected to either carbon atom 4 or carbon atom 6 in another glucose ring, 'α' defining the three-dimensional arrangement of the link.) Amylopectin in starch has branch sites at every 24 to 30 glucose residues. Glycogen branches more frequently, with branching at every 8th to 10th glucose unit.[4]

Extrafloral nectaries (*i.e.* structures that provide nectar but are not associated with flowers) also provide attractive food for ants in myrmecophytes. They may be glands on leaf blades, stems, or fruits and have varying shapes and sizes. Extrafloral nectaries serve as a source of nutritious sugars, especially glucose, fructose, and sucrose (Figure 2.14), as well as of water. The chemical compositions of their extrafloral nectars vary greatly with the type of plants.

Acacia trees (*Acacia* sp., Mimosaceae) are well-known for their extrafloral nectaries that attract ants. Ant-acacias (sometimes called 'swollen-thorn acacias') are found in Africa as well as in Central America. Acacia trees in the African savanna, another nutrient-poor environment, are notorious for their defensive ant inhabitants.

Acacias native to Central America have extrafloral nectaries as well as food bodies that offer nutrition for specialized ants. The acacia ant (*Pseudomyrmex ferruginea*) is a New World ant species that lives in a mutualistic relationship with the bullhorn acacia (*Acacia cornigera*). Figure 9.1(c) shows the large thorns of this acacia. The thorns are hollow, have an entry hole, and provide shelter for the *Pseudomyrmex* ants. Aside from extrafloral nectaries, Central American acacias also feature special food bodies that are at the tips of their leaflets (Figure 9.1(c)). Studies of the chemical composition of these food bodies have shown that they contain all the amino acids and all the fatty acids considered essential for insects and that they have a high content of lipids and proteins.[5] Therefore, they provide full nutrition for the inhabiting ants, without the need of external food sources.

9.3 Diverse Mutualistic Insect–Plant Interactions

Ever more examples of mutualism between plants and ants and other insects, often unusual and complex, are observed and investigated. The following two examples serve as illustrations.

The fanged pitcher plant (*Nepenthes bicalcarata*, Figure 9.3(a)), thus named because of its two large nectaries in the shape of fangs, is an endemic plant of Borneo. It is an insectivorous plant as well as a myrmecophyte, a seemingly contradictory situation. On one hand, the pitcher plant attracts insects to its water-filled pitchers and traps and digests them; on the other hand, it has a mutualistic relationship with ants that find shelter in its large tendrils attached to the pitchers. Specific ants (*Camponotus schmitzi*) have been found to regularly clean the rims of the pitchers that are the main surface for capture of prey. Enzymes in the water containers of these pitcher plants are of considerably lower acidity than in other pitcher plants. Thus, the specially-adapted ants can dive into the pitcher fluids and find food in the form of drowned insects. As for the benefit to the plants, it is widely believed that the ants prevent putrefaction from excess prey in the pitchers as well as keeping the insect-attracting rims effective. Within the *Nepenthes* genus, the fanged pitcher plant has exceptionally long-lived pitchers.[6]

(a)   (b)

Figure 9.3   Diverse plants with indirect plant defenses involving mutualistic ants. (a) Fanged pitcher plant (*Nepenthes bicalcarata*). (b) Serpentine columbine (*Aquilegia eximia*), with small insects trapped on the stems. (Photo by Carlo H. Séquin.)

Numerous and diverse herbaceous plants feature sticky glands on stems and leaves that contain glabrous exudates, often in the form of polysaccharides and sometimes containing terpenoids as well. (Compare Chapters 3.2 and 4.2.) These plants trap insects that eventually die on the sticky surfaces. Many of the plants have been suspected to also attract mutualistic insects, like ants, that provide indirect plant defense. The ants prey on the trapped insects and, at the same time, defend their territory from larger herbivores (and humans that may touch or try to weed out the plant). Such defense can serve as a form of protection for the respective plants. The serpentine columbine (*Aquilegia eximia*, Figure 9.3(b)), a plant native to California, has glands with sticky exudates on its stems that entrap small arthropods. The trapped insects attract ants that feed on the carrion. The presence of the ants indirectly discourages larger herbivores to browse on the plants. It is likely that sticky plants are commonly involved in such indirect defense strategies.[7]

9.4   Conclusions

Interactions that involve mutually beneficial interactions between insects and plants, especially with respect to defense of the plants, are increasingly discovered and investigated. They commonly involve several different organisms, in complex multipartite mechanisms.

Plants produce food bodies and nectaries with distinct compositions of nutrients that attract specific insects, especially ants. The insects in turn provide protective services for their host plants. There are selective advantages for plants that harbor these mutualistic insects, in processes that point to the coevolutionary relationships between plants and insects.

References

1. P. J. Gullan and P. S. Cranston, *The Insects: An Outline of Entomology*, Wiley Blackwell, Chichester, West Sussex, 4th edn, 2010, ch. 11.4, pp. 303–306.
2. P. Jolivet, *Interrelationship Between Insects and Plants*, CRC Press, Boca Raton, 1998.
3. S. Bischof, M. Umhang, S. Eicke, S. Streb, W. Qi and S. C. Zeeman, *Cecropia peltata* accumulates starch or soluble glycogen by differentially regulating starch biosynthetic genes, *Plant Cell*, 2013, **25**(4), 1400.
4. D. L. Nelson and M. M. Cox, *Lehninger Principles of Biochemistry*, W. H. Freeman, New York, NY, 6th edn, 2012.
5. M. Heil, B. Baumann, R. Krüger and K. E. Linsenmair, Main nutrient compounds in food bodies of Mexican *Acacia* ant-plants, *Chemoecology*, 2004, **14**, 45.
6. D. G. Thornham, J. M. Smith, T. U. Grafe and W. Federle, Setting the trap: cleaning behaviour of *Camponotus schmitzi* ants increases long-term capture efficiency of their pitcher plant host, Nepenthes bicalcarata, *Funct. Ecol.*, 2012, **26**, 11.
7. E. F. LoPresti, I. S. Pearse and G. K. Charles, The siren song of a sticky plant: Columbines provision mutualist arthropods by attracting and killing passerby insects, *Ecology*, 2015, **96**(11), 2862.

Part 3: Plants and Insects: The Human Perspective

10 Human Uses

10.1 Introduction

Many plant–insect interactions crucially affect human uses and needs. We depend on insects as pollinators of plants that are vital for human food production. On the other hand, insect infestations of crop plants, timber, and ornamental plants cause great damage and destroy vast acreages of essential plants. Some products of insect–plant interactions, like honey and beeswax, directly benefit our lives. Other important products are obtained from insects that have defined plant requirements. Silk is produced by silk moths whose larvae have to be raised on a specialized plant diet. As for some historic colorants, the dyes kermes and cochineal are derived from insects that feed on specific host plants. Furthermore, observations of plants that are avoided by insects have pointed to defensive plant compounds that could be used as insect repellents by people. As in previous chapters, the descriptions include discussions of chemical compounds and some reactions that are part of these relationships.

10.2 Essential Pollination for Food Production

Plants use a large number of strategies, many of them involving chemistry, to attract insect pollinators. Plants lure insects with scents, colors, nectar, and pollen to their flowers, as described in Chapter 2. In return, insects transfer pollen onto the stigmas of flowers and induce fertilization. Pollination by insects is of central importance for the fertilization of crop plants.

The Chemistry of Plants and Insects: Plants, Bugs, and Molecules
By Margareta Séquin
© Margareta Séquin 2017
Published by the Royal Society of Chemistry, www.rsc.org

The large majority of crops raised for human consumption depend on animal pollination. Various types of bees are the main pollinators, pollinating about 75% of cultivated plants worldwide (with a higher percentage reported for Europe and other temperate zones). Bee-pollinated plants include orchard crops like apples, almonds, and citrus, as well as cotton, coffee, soybeans, cabbage, and potatoes. Flies and bees pollinate plants like onion, lettuce, peppers, and yam. Wasps, the pollinators of figs, constitute about 5% of pollinating animals of food plants. Beetles are required for the pollination of oil palms. Butterflies and moths visit various crop plants and pollinate about 4% of cultivated plants. The remaining animal-pollinated crops are pollinated by bats and birds. While wind-pollinated staple foods, like cereals, rice, and maize, make up about 60% of the production volumes of principal crops, a much greater diversity of plant species are pollinated by animals: out of 115 principal world crops that provide food for humans, 87 require animal pollination, mostly by bees, compared to 28 crop plant species that are pollinated by wind.[1,2] The above summary indicates the great importance of bee pollinators for human food production.

There are more than 16 000 different known species of bees.[3] It is likely that many more types of bees exist, but have not yet been discovered. Bees live on every continent, except Antarctica, and in every habitat that has insect-pollinated flowering plants. While some bees live in colonies, others are solitary. Bees are very efficient pollinators. They are especially adapted to carrying pollen and visit flowers more often than any other insects or animals in general. Pollen provides food not only for the bees themselves but also for their broods. Honey bees (*Apis mellifera*) have an excellent communication system, communicating to other bees in the hive the locations of plants to visit for pollen and nectar. Bees tend to visit one type of flower or just a few related flower species (Figure 10.1(a) and (b)). Many bees have environmentally and thermally well-controlled nests and therefore adapt better to changing environmental conditions, like a frost or a heat wave, than other insects.

Bees other than honey bees are of great importance as pollinators as well. Solitary bees like *Andrena* are active in early spring and thus are important for pollinating early-flowering orchard plants. Bumble bees, with their larger size, can carry more pollen and can do more pollinations per visit than honey bees. Because of their hairiness, bumble bees can be active in cold weather, when pollen in many plants is at its peak. Therefore, maintaining the diversity of different species of bees is of crucial benefit for our crops – aside from supporting the natural biodiversity.

(a)   (b)

Figure 10.1 Pollination by bees. (a) Honey bee (*Apis mellifera*) pollinating blackberry blossoms. (b) Honey bee, with pollen sacs on hind legs, on mustard plant.

10.3 Honey

Several types of bees produce and store honey, including some bumble bees. But only species of the genus *Apis* build hives with extensive honey accumulation and thus are tended for honey production for human consumption. We focus here on the most common producer of honey, namely the Western honey bee (*Apis mellifera*, Figures 10.1(a) and (b) and 10.2(a)).

The principal raw material for honey is nectar which the honey bees suck up from flowers. Nectar is mostly a dilute solution of sugars dissolved in water. (See Chapter 2.8 for details on the composition of floral nectars.) Flower nectars contain the disaccharide sucrose as well as the monosaccharides glucose and fructose. Chemical transformations of the nectar components begin in the bee's digestive system immediately after collection. They particularly involve the breakdown of sucrose **10.1** into glucose **10.2** and fructose **10.3**, by the action of the enzymes diastase and invertase in the bee's stomach (Figure 10.3). In addition, the enzyme glucose oxidase catalyzes the oxidation of a portion of glucose, *e.g.* into gluconic acid **10.4**. Bees have a honey stomach which is a widened region at the bottom of their esophagus.[4] The foraging bee transfers the nectar, that is already undergoing chemical changes, to a worker bee in the hive. Eventually the nectar droplet is deposited in a cell in the honeycomb (Figure 10.2(a)). The transformation of nectar into honey involves two major processes: the evaporation of excess water and the conversion of sucrose into glucose and fructose. Bees are constantly fanning the honeycombs with their wings in order to speed up water evaporation, to less than 20% water content. The conversion from collected nectar to honey, called

Figure 10.2 Honey bees and honey. (a) Honey bees in the hive. (b) Honey samples, liquid or granulated, from different floral sources have different flavors and colors.

'ripening', can take about three weeks. When a cell in the honeycomb is full, bees cap the cell with newly produced beeswax.[5]

Bees also gather pollen into pollen baskets on their hind legs, as can be seen in Figure 10.1(b). Pollen carried on the bees' bodies may be carried to another flower where a small portion rubs off onto the flower's pistil and induces cross pollination. Bees transport pollen back to the hive where it is used as food for the adult bees and for the developing larvae. It is an important source of proteins for them. Honey always contains some pollen.

In preparation for human consumption, beekeepers generally remove honey from the honeycombs by spinning them in a centrifuge to separate liquid honey from solid wax. The honey is then heated to destroy sugar-fermenting yeasts, commonly followed by filtration to remove any remaining wax or debris. Different honey samples may be blended. Honey samples from different floral sources have different flavors and colors, from light yellow to darker colors (Figure 10.2(b)). Liquid honey tends to granulate upon storage, especially at lower temperatures. It can be made liquid again with gentle heat.[6]

Honey is a viscous, sweet food which humans have used and enjoyed as a natural sweetener for thousands of years. It is a supersaturated solution of sugars, mainly fructose and glucose. The fructose content in honey samples from the USA is listed as 38.4% on average, with an average glucose concentration of 30.3%. The origin of the honey, the level of maturity of the sample, and the processing and storage conditions lead to wide ranges of the components' concentrations (see Table 10.1). This includes the water content of honey which can range from 12.2 to 22.9%.[7]

Table 10.1 Chemical Composition of U. S. Honeys.[a]

Component	Average (%)	Range (%)
Water	17.2	12.2–22.9
Fructose	38.4	30.9–44.3
Glucose	30.3	22.9–40.7
Sucrose	1.3	0.2–7.6
Gluconic acid	0.57	0.17–1.17
Other acids	0.43	0.13–0.92
Minerals	0.17	0.02–1.03

[a]The pH of honey is 3.9 on average and has a range of about 3.4 to 6.

Honey is slightly acidic and has an average pH of about 3.9. The acidity is the result of organic acids that are formed through enzymatic oxidation of reducing sugars, like glucose, when nectar is transformed into honey. Gluconic acid **10.4**, an oxidation product of glucose, is the most common of the organic acids in honey (Figure 10.3). The natural acidity of honey inhibits the growth of microorganisms in it. In addition, bacteria cannot survive the osmotic pressure of a sugar solution as highly concentrated as honey. The combination of a low pH, a high sugar concentration, and a low water content make honey a very stable product, with a long shelf life.

We pause here our discussions of honey compositions to take a look at different representations of carbohydrate structures that are commonly used to describe mono- and disaccharides. Carbohydrates have complex structures, with numerous chiral centers. To present their molecules in an informative manner, several ways of illustrating them are used, depending on which aspects need to be discussed. So far we have shown them as Haworth projections (designed by Sir W. N. Haworth, Nobel Prize 1937). Sucrose **10.1**, glucose **10.2**, and fructose **10.3** are presented as Haworth projections in Figure 10.3 (see also Chapter 2.8). These common illustrations of carbohydrate structures are simplified pictures of the actual chair shapes, or chair conformations, of the molecules, as shown for glucose below its Haworth projections in Figure 10.3. Glucose and fructose molecules not only exist in ring forms, but also as open chains. They form equilibria between open-chain and ring forms, leading to α- and β-forms as shown in Figure 10.3. Open-chain structures are commonly drawn as Fischer projections (H. E. Fischer, Nobel Prize 1902). The open-chain form of glucose **10.2a** is shown as a Fischer projection in Figure 10.3. Its enzymatic oxidation forms gluconic acid **10.4** and hydrogen peroxide (H_2O_2).[8]

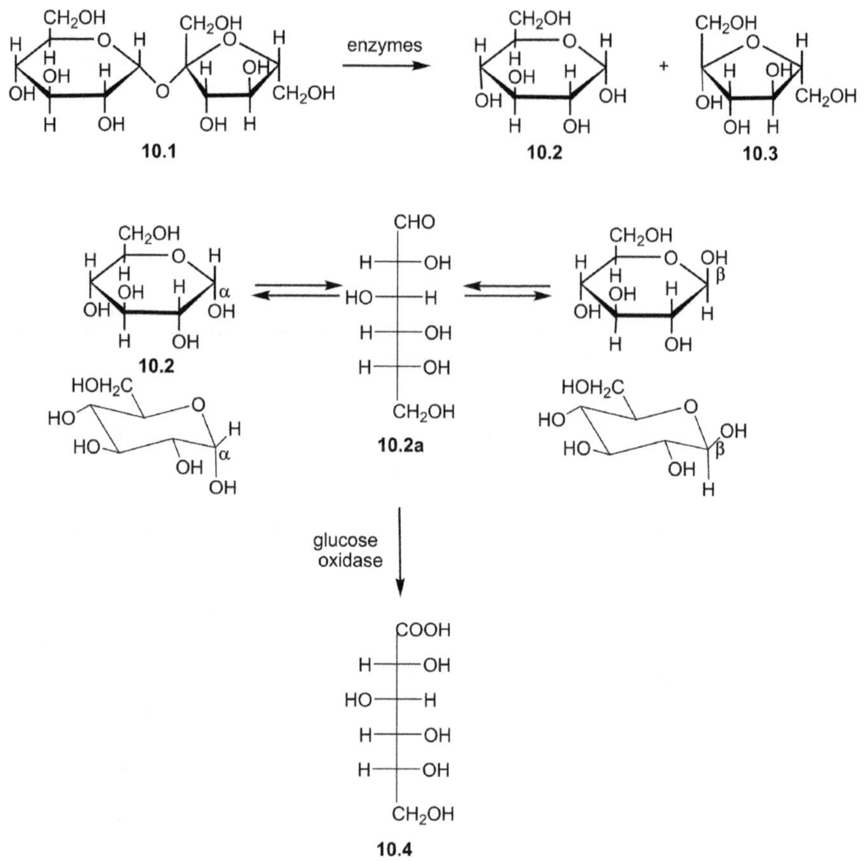

Figure 10.3 Sucrose **10.1** from nectar breaks down enzymatically into glucose **10.2** and fructose **10.3** in bees' honey stomachs. The open-chain form of glucose **10.2a** is in equilibrium with its α- and β-ring forms, shown as Haworth projections and as chair forms. Oxidation of glucose forms gluconic acid **10.4**.

Aside from sugars, water, and organic acids, honey contains amino acids, proteins, and some vitamins. Minerals like potassium, calcium, and sodium, as well as phenolic compounds are also found. Pollen is the main source of proteins and amino acids, with proline **2.24** (shown in Chapter 2) being the most abundant amino acid in honey. Only small amounts of vitamins are present; they are mostly B vitamins and vitamin C and are derived from the pollen grains suspended in honey. Some honey types, like the dark manuka honey from New Zealand, have a high phenolic content due to tannins. Phenolics, hydrogen peroxide from the oxidation of glucose, and a low pH give honey antibacterial properties that may contribute to reported beneficial health effects. Honey has been used in wound dressings since ancient times.[9]

The colors, aromas, and flavors of honey samples vary greatly and depend on the floral origins of the nectars. Bees tend to forage on one or just a few types of flowers and their nectars, which results in honeys with characteristic flavors and colors. Honey that is derived from clover tastes different from a sample of honey from orange blossoms or lavender. More than two hundred volatile compounds have been identified from different honey samples as part of their typical aromas and fragrances. Among the volatiles are aliphatic aldehydes and ketones, alcohols, esters of aliphatic and aromatic acids, as well as monoterpenes. Figure 10.4 shows a few examples of aroma compounds and volatiles detected in honey samples. Vanillic acid **10.6** is a common phenolic in honey, contributing to its pleasant flavor. Methyl anthranilate **10.7** is a typical flavor component of citrus and lavender honey; the compound has a fruity, grapy note. Monoterpenes, like *cis*-rose oxide **10.8**, with a rose scent, are common in many floral honeys. Volatiles may evaporate over time, which results in a change of the composition of honey – and with this its taste – with long storage.

Pigments give honey samples their color, with light to darker colorations mostly determined by the sources of the floral nectars (Figure 10.2(b)). Long storage of honey or high heat leads to darkening. This change of color may be accompanied by the formation of off-flavored products from a series of reactions known as Maillard reactions (according to L.-C. Maillard, 1912). These chemical reactions occur between amino groups (*e.g.* from amino acids) and reducing sugars like glucose. They occur especially at higher temperatures, with the product structures undergoing many rearrangements. One product is 5-hydroxymethylfurfural **10.9** (Figure 10.5), formed by the decomposition of monosaccharides. Its concentration

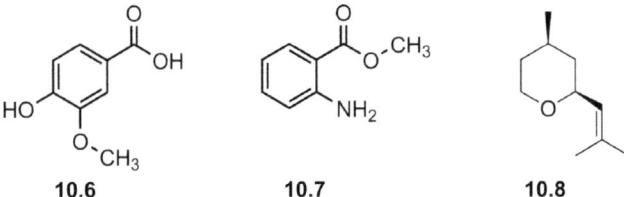

| 10.6 | 10.7 | 10.8 |

Figure 10.4 Flavor components in honey. Vanillic acid **10.6**, methyl anthranilate **10.7**, and the monoterpene *cis*-rose oxide **10.8** contribute to typical flavors of honey samples.

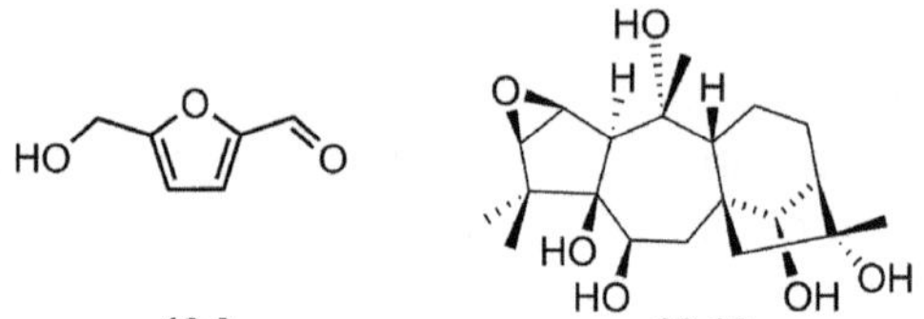

Figure 10.5   Undesirable honey components. 5-Hydroxymethylfurfural **10.9** is increasingly formed when honey is heated or upon long storage. Grayanotoxins, like **10.10**, are isoprenoid toxins from *Rhododendron* sp.

increases with high-heat treatment of honey and long storage; it is an indicator of the age of a sample. Other degradation products include nitrogenous organic compounds with distinct flavors. These reactions affect taste and odor of the honey. (The products of Maillard reactions and the following arrangements and their products are well known to flavor chemists. They are the typical flavorful compounds known from toasted, fried, or baked goods and meats, the result of sugars reacting with amino acids upon heating.)[10]

As nectar from flowers is the source of honey, nectar toxins may also be found in honey, the famous – or infamous – example being honey from rhododendrons and other plants of the Ericaceae. Called 'mad honey', it is especially known from the Eastern Black Sea region. Its effect on people has been described as early as 401 BC by Xenophon, a military commander and author from Athens.[11] Rhododendron nectar and the honey derived from it contain grayanotoxins, a group of toxic diterpenes (C_{20}) with complex structures.[12] Figure 10.5 shows a grayanotoxin **10.10** from *Rhododendron japonicum.* (Can you spot the isoprene units in its structure? For review check Chapter 2.6.) Bees are not affected by their toxicity, but humans experience ill-effects such as dizziness or nausea.

Honeys from single types of flowers, with distinct flavors and sometimes with specific attributed health effects, can be sold for higher prices than those from multiple floral sources. This has led to the production of dilutions with honey of lower value and to fraudulent mislabeling. A high sucrose content, commonly accomplished by adding sucrose from sugar beets, may point to adulteration. Several techniques have been developed in order to detect adulteration of honey samples. The honey's origins can be studied by examining its pollen grains under a microscope, as pollen grains are unique to each type of flower. But this is a tedious, time-consuming method. A faster way to analyze honey samples has been recently developed; it uses NMR (nuclear magnetic resonance) spectroscopy to produce characteristic fingerprints of components in the extracts. NMR spectroscopy is a most important technique that chemists and biochemists use to determine structures of organic molecules.[13] Among its many applications, this spectroscopic method allows to determine the number, the kind, and the relative locations of atoms in an organic molecule. Very commonly, NMR focuses on the hydrogen atoms in a molecule (H-NMR), but also on carbon atoms (and a few others). The authors that developed the NMR method to test chloroform extracts from almost 1000 monofloral honey samples, of different botanical origins, then examined the samples by H-NMR to determine characteristic hydrogen signals. Volatile compounds from flowers are relatively small organic molecules, with not too many carbon and hydrogen atoms composing them (see Chapter 2.3). As such they produce distinct, well-separated hydrogen signals in an H-NMR spectrum. Extracts from honey samples were found to produce spectra that serve as 'fingerprints'; they represent characteristic hydrogen signals that can be assigned to the different flavor compounds. Comparison of unknown honey samples with standard monofloral samples makes it possible to determine the identities and the number of floral origins of honey samples in an efficient manner.[14]

10.4 Beeswax

Worker honey bees (*Apis mellifera*) can produce wax from special glands on their abdomens. The freshly secreted wax is colorless. Gums and resins, as well as fragrances from the floral sources the bees were visiting, are mixed in by the bees. They use the secretions to form cells in which they store honey and protect their larvae in the beehive. Beekeepers use a variety of methods to render beeswax from combs, such as applying steam, or using presses or centrifugation. The beeswax product has a yellowish coloration and features a pleasant, honey-like fragrance (Figure 10.6(a)).

Beeswax has been used by humans since ancient times. It was used in embalming Egyptian mummies, in ship-building, as well as an ingredient in ancient paintings and other art objects, and for making wax molds. Beeswax has long been a most important material for candle-making. In modern times, beeswax continues to have many applications. It is an ingredient in cosmetics, in pharmaceuticals, as well as a component in coatings and polishes; it is still

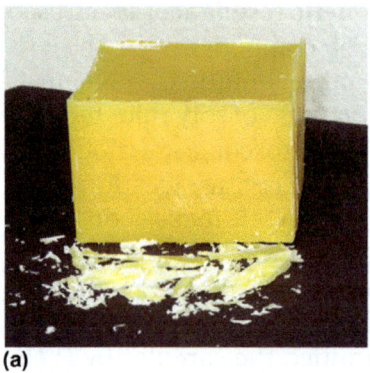
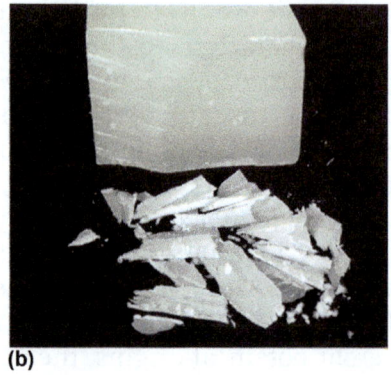

(a)   **(b)**

Figure 10.6   Beeswax and paraffin wax. (a) A block of natural beeswax. (b) Paraffin wax obtained from petroleum.

used in candle-making. Its high stability and longevity have made it a sought-after material all through history.[15]

Beeswax has a melting point around 60 °C, with a range from 62 °C to 64 °C. It is solid and brittle at low temperatures and softens when held in the hands. The chemical composition of beeswax is complex. Its main constituents are esters of fatty acids and long-chain alcohols (typically with 30–32 carbon atoms). Figure 10.7 shows the C_{16} fatty acid palmitic acid **10.11**, a common saturated fatty acid in biological waxes, and 1-triacontanol **10.12**, a C_{30} alcohol. Their ester, triacontanyl palmitate **10.13**, is the major component in beeswax. Minor components are fatty acids, long-chain alcohols, other esters, and some long-chain aliphatic hydrocarbons. Sterols, like cholesterol, and fragrant compounds in the form of terpenes, like geraniol, are also part of the mixture.

Note that all the components of beeswax are water-repelling compounds. Their long carbon chains point to their hydrophobic (and lipophilic) properties. Waxes in general, whether from biological or synthetic sources, are water-repelling substances that are pliable and soften upon warming. They are harder and less greasy than fats and oils. The chemical structures of wax components always include long hydrocarbon chains. Biological waxes are generally mixtures of esters of long-chain alcohols with fatty acids, with minor percentages of other lipophilic compounds, like sterols and long-chain alcohols. These waxes act as protective coatings on leaves and fruits of many plants. They also have insulating functions on many animal surfaces, like outer surfaces of insects, on skins, or on feathers of birds.

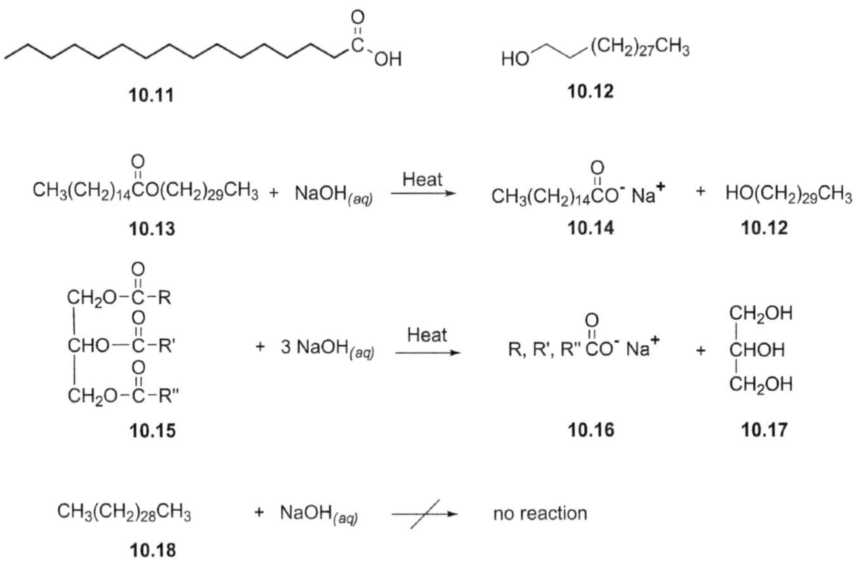

Figure 10.7 Beeswax, fats and oils, and synthetic paraffin wax: a comparison. The esters triacontanyl palmitate **10.13** and fats and oils (triglycerides) **10.15** react with sodium hydroxide solution to form soaps (**10.14** and **10.16**), unlike the paraffin triacontane **10.18**, a component of paraffin wax, which is inert towards treatment with base.

Beeswax, due to its major composition of esters and minor components of fatty acids, will react with aqueous sodium hydroxide solution to form soap, *i.e.* sodium salts of fatty acids **10.14**. While waxes are mostly simple esters, fats and oils, in comparison, have more complex structures. They are triesters of glycerol. Figure 10.7 shows the general structure of a fat or oil **10.15**, namely a triglyceride, with the different hydrocarbon groups from the fatty acids labeled as R, R′, and R″. Fats and oils are saponified, *i.e.* hydrolyzed when reacted with aqueous NaOH solution, in the well-known reactions that produce their sodium salts **10.16**, *i.e.* soap, and glycerol **10.17**.

In contrast, paraffin waxes (Figure 10.6(b)), obtained from crude oil, consist of long-chain alkanes (also known as paraffins), like triacontane, $C_{30}H_{62}$ **10.18**. The name 'paraffin' is derived from 'parum affinis', Latin for 'lacking affinity or reactivity'. It signifies that these compounds are unreactive and show no reaction with aqueous NaOH, unlike the reactivity of biological waxes or of fats and oils towards solutions of strong base. Paraffin waxes have waxy properties and have many applications in products for lubrication or insulation, as well as in candle-making.

10.5 Silk

One of the most precious materials produced by insects is silk from the cocoons of silkworms. Beginning in China, silk moths (*Bombyx mori*) and their larvae have been domesticated and raised for around 5000 years for the exquisite silk thread that the caterpillars spin when they form a cocoon (Figure 10.8(a) and (b)). Male silkworms produce the preferred silk, as a single thread of 300 to 900 m per cocoon. Silkworms, *i.e.* the larvae or caterpillars of silk moths, have specific plant food requirements. They must be fed mulberry leaves, preferably of the white mulberry tree (*Morus alba*). The larvae, in an example of monophagy, reject other plant food. Female silk moths (Figure 5.5(a)) are somewhat larger than males and attract males with minute quantities of the pheromone bombykol **5.5** (Figure 5.6). Females are needed for mating and to lay eggs, but produce cocoons of lower quality. Through domestication, male silk moths have lost their ability to fly. To obtain the silk thread for further processing, silkworm cocoons are boiled which kills the larvae. Silk is produced by other arthropods as well, like spiders, but only silk from silk moths has been used for manufacturing textiles.[16]

Silk has outstanding properties that set it apart from all other natural fibers. It is the strongest natural fiber and has high extensibility, *i.e.* can be stretched extensively without tearing. These

(a)   (b)

Figure 10.8   Silkworm and cocoon. (a) Caterpillar of *Bombyx mori* feeding on mulberry leaves (*Morus alba*). (b) Cocoon, about 2 cm in length, of male silkworm on silk fabric.

combined properties make silk tougher than any synthetic fiber available today. Silk has a high capacity to pick up moisture, up to about 30% of its weight. Silk fabrics have a smooth, lustrous texture which has long made them highly desirable as textiles and as materials for works of art. Its exclusive properties historically led to extensive trading. Silk and many other goods, like spices and dyes, were transported from the Orient along the famous Silk Road, an ancient network of trade routes. Because of its great strength and its biocompatibility, silk has important modern uses in the medical field, *e.g.* as sutures in surgery. As silk is an important commodity and as silkworms are easy to obtain and to raise, extensive research has been carried out relating to the structure and the production of silk.

Silk, an animal fiber, consists of proteins. Its two main proteins are fibroin and sericin. Fibroin contributes 70–80% of the proteins and provides the structure of silk. Recall that proteins are polymers of amino acid sequences and have defined three-dimensional shapes (Chapter 2.10 and Figure 2.24.). Fibroin is an example of a fibrous protein, with an elongated shape. It consists of amino acid sequences that contain numerous repeated glycine and alanine monomers. Glycine **10.19** and L-alanine **10.20** are the two smallest amino acids among those that compose natural proteins (Figure 10.9). Note that glycine has no chiral center. The high content of these strongly polar amino acids with few carbons, leads to extensive hydrogen bonding between the protein strands. This in turn results in dense packing of the polymer strands, forming a strong fiber. It also explains the high absorption rate of moisture in silk.[17] The second silk protein is sericin, providing 20–30% of the silk fiber. Sericin is a glycoprotein, which means that it has carbohydrate chains ('glyco') attached to sidechains of the amino acid sequences. It is a globular protein, with a roundish shape, and acts as a sticky coating between the fibers. Sericin had long been considered a waste product in silk production, being routinely washed out from the silk fibers to obtain the luster of textiles from fibroin alone. It has recently been found that sericin can have great uses in biomedical applications. Because it picks up

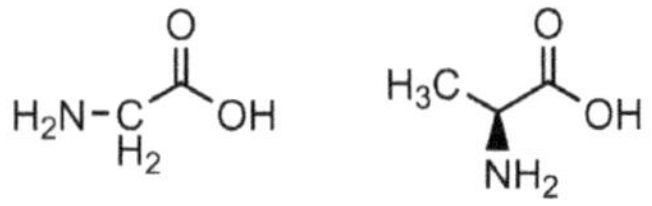

10.19   10.20

Figure 10.9   The amino acids glycine **10.19** and L-alanine **10.20** are the main monomers of fibroin in silk.

moisture easily and is biocompatible, sericin can be useful in tissue engineering, *e.g.* for skin tissue repair as part of wound dressings.[18] In the glands of the spinning mechanisms of silkworms the protein mixture is in aqueous solution, in concentrations higher than 30%. These are concentrations in which most proteins become water-insoluble and solidify, a process that would block the spinning apparatus of the caterpillars. The mechanisms that allow the silk proteins to achieve metastable states that keep them in solution during the spinning of the cocoon are still unknown.[19]

We conclude with a look at the highly specialized food requirements of domesticated silk worms, namely their need to feed on mulberry leaves (*Morus alba*). The combination of several secondary plant metabolites in the leaves have been found to provide the attractiveness of this plant food to the insects (Figure 10.10). Several terpenes act as olfactory attractants and appeal to the caterpillars' sense of smell. Among them is citral, composed of a 2 : 1 mixture of geranial **10.21** and its isomer neral **10.22**. Both compounds are volatile, fragrant monoterpenes. While these terpenes are common in plants, two other compounds, namely morin **10.23** and the

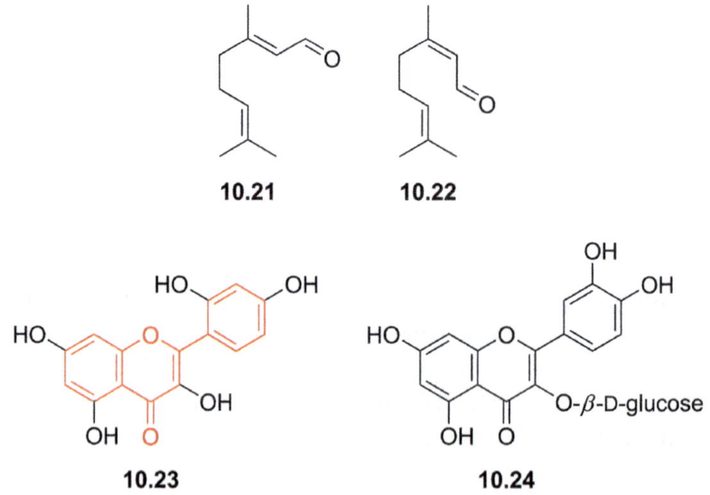

Figure 10.10   Chemical silkworm attractants in *Morus alba*. Citral, a 2 : 1 mixture of the monoterpenes geranial **10.21** and its isomer neral **10.22**, acts as an attractant for the caterpillars' olfactory system. The flavonoids morin **10.23** and isoquercitrin **10.24** are 'biting factors'. (The characteristic three-ring system of flavonoids is highlighted in **10.23**.)

glucoside isoquercitrin **10.24**, are restricted in their occurrence and are typical components of mulberry leaves. They are not a general dietary requirement for insects, but they make the actual feeding attractive to the silkworms and have been cited as 'biting factors'. Morin and isoquercitrin are flavonoids and feature the typical three-ring system of flavones (highlighted in **10.23**). Note the different substitution patterns and the glucoside attachment in isoquercitrin. Attempts to alter the silkworms' food of mulberry leaves, like replacing it partially with synthetic diets, have been found to have a poor influence on their feeding desire and on the quality of the produced silk fibers.[20,21]

10.6   Dyes From Insects

Two ancient dyes have their sources in insects: kermes and cochineal. Kermes, a red dye, is derived from the kermes scale insect *Kermes vermilio* occurring on kermes oak trees in the Middle East and the Mediterranean region. (The word 'crimson' for a deep-red color is derived from 'kermes'.) Cochineal is a red dye obtained from the cochineal scale insect (*Dactylopius coccus*) that lives on cacti of the southwest of North America and in Mexico. The colorful solutions produced by the insects are a form of defense towards predators. Both colorants had ancient uses and were extensively used during the Middle Ages as they are excellent dyes for protein fibers like wool and silk. Cochineal, more permanent as a dye than kermes, is still used in carpet making and in art work. It has applications in cosmetics and as a food additive.

Kermes and cochineal insects have specific plant requirements. Kermes insects specifically feed on the sap of kermes oaks (*Quercus coccifera*), trees of the Mediterranean region and North Africa (Figure 10.11(a)). Cochineal insects live on cactus pads of the genus *Opuntia* (Figure 10.11(b)) where they ingest sap from the cactus pads.

The chemical structures of the dyes are closely related to each other. Kermesic acid **10.25** from kermes insects and carminic acid **10.26** from cochineal insects are anthraquinones (Figure 10.12). (The anthraquinone structure is highlighted in kermesic acid.) Carminic acid is a *C*-glucoside of kermesic acid. Anthraquinone dyes are also a large branch of synthetic dyes. Their structures have been inspired by the importance and desirable qualities of the natural colorants.[22,23]

Figure 10.11 Host plants of scale insects, sources of ancient dyes. (a) Kermes oak (*Quercus coccifera*), host plant of the scale insect *Kermes vermilio*. (Photo by Wikimedia Commons. https://commons.wikimedia.org/wiki/File:Quercus_coccifera_ %28Kermes_Oak%29,_Agiasos,_Lesbos,_Greece.jpg, (accessed: October 2016)). (b) Cochineal scale insects (*Dactylopius coccus*) on prickly pear cactus (*Opuntia* sp.).

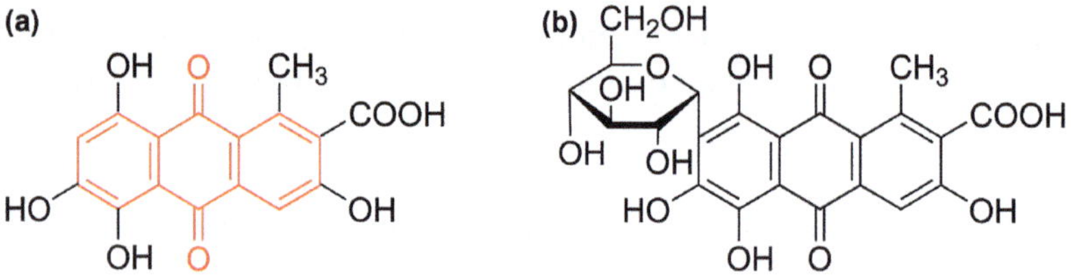

10.25

10.26

Figure 10.12 Ancient dyes of carmine color: kermes and cochineal. (a) Kermesic acid **10.25**, aglycone of carminic acid. (The anthraquinone structure is highlighted.) (b) Carminic acid **10.26**, α-C-glucoside of **10.25**.

10.7 Insecticides From Plants

People have long observed that certain plants are avoided by herbivorous insects. Plants have evolved a large number of diverse defensive compounds that repel attacking insects (see Chapter 4). An example of such a plant is the neem tree (*Azadirachta indica*, described in Chapter 4.6) which is shunned by insects. Its leaves, fruits, and seeds contain the bitter, insect-repelling terpenoid azadirachtin **4.25**. With large swaths of food crops regularly falling prey to insect infestations, the search is on to find effective methods that prevent or minimize the damage. Plant compounds that repel or kill herbivores provide inspiration, especially for the development of insecticides that are biodegradable and that have low toxicity towards higher animals. The following examples describe insect-repelling plant compounds that do not only have uses as insecticides themselves, but that have also led to the development of synthetic derivatives for use in managing insect pests. Natural plant-derived insect repellents are sometimes referred to as 'botanicals' (which does not imply any information about their harmlessness or toxicity).

The pyrethrum plant (*Tanacetum cinerariifolium*, formerly *Chrysanthemum cinerariifolium*) is a common plant with daisy-like flowers, originating from the Mediterranean region (Figure 10.13(a)). For centuries, extracts from dried flower heads of pyrethrum have been applied to clothes, as well as to cloth bags holding grains, in order to kill unwanted insects like fleas or weevils. The effectiveness of pyrethrum, aside from its low toxicity towards higher animals, made it an important commodity for trading as well.[24]

Pyrethrum plants contain a group of related compounds collectively known as pyrethrins. The elucidation of their chemical structures was accomplished by chemists H. Staudinger and L. Ružička in 1924, long

(a) (b) (c)

Figure 10.13 Plants that contain potent natural insecticides. (a) Pyrethrum (*Tanacetum cinerariifolium, Chrysanthemum cinerariifolium*), a source of pyrethrins. (b) Poison vine (*Derris elliptica*, Fabaceae), a source of rotenone. (Photo by Forest & Kim Starr. https://upload.wikimedia.org/wikipedia/commons/5/5c/Starr_010425-0043_Derris_elliptica.jpg, (accessed October 2016)). (c) Tobacco plant (*Nicotiana tabacum*), source of nicotine.

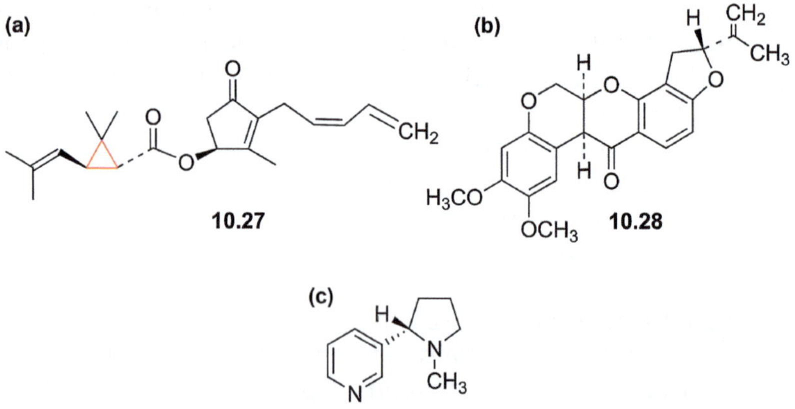

Figure 10.14   Natural insecticides from plants. (a) Pyrethrin I **10.27** is one of the naturally occurring insecticidal terpenoids in pyrethrum flowers. (The cyclopropane ring in its structure is highlighted.) (b) Rotenone **10.28** occurs in *Derris* sp. and other plants of the Fabaceae. (c) Nicotine **10.29** from *Nicotiana* sp. is a highly toxic alkaloid with insecticidal properties.

after the plants began being used as insecticides.[25] Figure 10.14 shows the structure of pyrethrin I **10.27**. Pyrethrins are esters (COOR) and feature a cyclopropane (C_3) ring (highlighted in **10.27**) which is uncommon in plant metabolites. They are terpenoids, *i.e.* related to terpenes. Their structures are biochemically assembled in pathways similar to terpenes, but do not contain a regular pattern of isoprene units. While pyrethrins repel insects at low doses, they are effective insecticides at higher doses, acting on the insects' nervous system as neurotoxins.

Pyrethrins have inspired a large group of synthetic insect repellents and insecticides because they are biodegradable, breaking down when exposed to light and oxygen, aside from their insecticidal potency. Once the structures of the natural compounds were determined, synthetic insecticides derived from them could be prepared. Their chemical structures strive for higher efficiency as insecticides, while retaining their biodegradability and low toxicity towards higher animals. Compounds that are synthetically prepared and close in structure to pyrethrins are known as 'pyrethroids'. A concern is the toxicity of pyrethrins and pyrethroids towards fish and other aquatic organisms.

Rotenone **10.28** is another example of a potent insecticidal compound from plants. It occurs in the roots of *Derris* sp. (Fabaceae,

Figure 10.13(b)), a vine from Southeast Asia that is also known as poison vine or tuba root. Rotenone is a broad-spectrum pesticide, killing insects unselectively, as well as fish, but has low toxicity towards mammals. The dried roots of *Derris* were traditionally used to catch fish by native people. Another source of rotenone is the Mexican plant jicama (*Pachyrhizus erosus*, Fabaceae) whose seeds contain the insecticide. Due to its broad-spectrum toxicity its use is gradually being phased out. The chemical structure of rotenone **10.28** (Figure 10.14) is distinctly different from the pyrethrins, featuring a multicyclic system related to flavonoids.

Nicotine **10.29**, an alkaloid from tobacco plants (*Nicotiana* sp., Figure 10.13(c)), is another potent broad-spectrum insecticide. (The compound was introduced in Chapter 4.7.) Extracts from tobacco have long been used as pesticides. Due to its high toxicity to mammals, nicotine has been phased out as an insecticide. But synthetic insecticides, namely the neonicotinoids, have been derived from the naturally occurring alkaloid. (Read more about these synthetic derivatives in the next chapter.)

The differences in the chemical structures of these examples of insecticides from plants are striking. The natural compounds provided guidelines for the development of synthetic products in the ongoing challenge to find products that are effective against unwanted insect pests, but that have a low toxicity towards the surrounding environment.[26]

10.8 Conclusions

Many plant–insect interactions greatly influence human lives. The beginning of this chapter made a case for honey bees and their vital role in the production of an adequate food supply for humans. Products like honey, beeswax, silk, and even dyes have long historic backgrounds, with honey, beeswax, and silk still being commodities of great importance today. The natural compounds and compositions have provided inspiration for the creation of new materials, in search of products that have more desirable properties, or that may be environmentally friendlier, an ongoing challenge for a changing world with a growing human population.

The next chapter will elaborate on different approaches to manage invasive herbivorous insects that at times infest and greatly damage plants in agriculture and horticulture.

References

1. P. Willmer, *Pollination and Floral Ecology*, Princeton University Press, Princeton, 2011.
2. I. H. Williams, The dependence of crop production within the European Union on pollination by honey bees, *Agric. Zool. Rev.*, 1994, **6**, 229.
3. B. N. Danforth, S. Sipes, J. Fang and S. G. Brady, The history of early bee diversification based on five genes plus morphology, *Proc. Natl. Acad. Sci. U. S. A.*, 2006, **103**(41), 15118.
4. R. J. Elzinga, *Fundamentals of Entomology*, Pearson, Upper Saddle River, NJ, 6th edn, 2004.
5. D. W. Ball, The Chemical Composition of Honey, *J. Chem. Educ.*, 2007, **84**(10), 1643.
6. H. McGee, *On Food and Cooking*, Scribner, New York, NY, 2004.
7. H.-D. Belitz, W. Grosch and P. Schieberle, *Food Chemistry*, Springer, Berlin, 4th edn, 2009.
8. K. P. C. Vollhardt and N. E. Schore, *Organic Chemistry: Structure and Function*, W. H. Freeman, New York, NY, 7th edn, 2014.
9. A. B. Jull, N. Walker and S. Deshpande, *Honey as a topical treatment for wounds*, The Cochrane Collaboration, Wiley & Sons, Ltd, 2014.
10. S. Everts, The Maillard Reaction Turns 100, *Chem. Eng. News*, 2012, **90**(40), 58.
11. A. Gunduz, S. Turedi, R. M. Russell and F. A. Ayaz, Clinical review of grayanotoxin/mad honey poisoning past and present, *Clin. Toxicol.*, 2008, **46**, 437.
12. E. Breitmaier, *Terpenes: Flavors, Fragrances, Pharmaca, Pheromones*, Wiley-VCH, Weinheim, 2006.
13. K. L. Williamson and K. M. Masters, *Macroscale and Microscale Organic Experiments*, Brooks/Cole, Belmont, CA, 6th edn, 2011.
14. E. Schievano, C. Finotello, J. Uddin, S. Mammi and L. Piana, Objective Definition of Monofloral and Polyfloral Honeys Based on NMR Metabolomic Profiling, *J. Agric. Food Chem.*, 2016, **64**(18), 3645.
15. R. A. Morse and W. L. Coggshall, *Beeswax: Production, Harvesting, Processing and Products*, Wicwas Press, Cheshire, CT, 1984.
16. G. Waldbauer, *Fireflies, Honey, and Silk*, Univ. of California Press, Berkeley, CA, 2009.
17. D. L. Nelson, M. M. Cox and A. L. Lehninger, *Lehninger Principles of Biochemistry*, W. H. Freeman, 6th edn, New York, NY, 2013.

18. L. Lamboni, M. Gauthier, G. Yang and Q. Wang, Silk sericin: A versatile material for tissue engineering and drug delivery, *Biotechnol. Adv.*, 2015, **33**(8), 1855.

19. F. G. Omenetto and D. L. Kaplan, New Opportunities for an Ancient Material, *Science*, 2010, **329**(5991), 528.

20. J. B. Harborne, *Introduction to Ecological Biochemistry*, Academic Press, London, 4th edn, 1993.

21. P. Jolivet, *Interrelationship Between Insects and Plants*, CRC Press, Boca Raton, 1998.

22. M. Séquin, *The Chemistry of Plants: Perfumes, Pigments, and Poisons*, Royal Society of Chemistry, Cambridge, UK, 2012.

23. M. Séquin-Frey, The chemistry of plant and animal dyes, *J. Chem. Educ.*, 1981, **58**, 301.

24. *Pyrethrum, the Natural Insecticide*, ed. J. E. Casida, Academic Press, New York, 1973.

25. H. Staudinger and L. Ruzicka, Insektentötende Stoffe I. Über Isolierung und Konstitution des wirksamen Teiles des dalmatinischen Insektenpulvers, *Helv. Chim. Acta*, 1924, **7**(1), 177.

26. S. Manahan, *Fundamentals of Environmental and Toxicological Chemistry: Sustainable Science*, CRC Press/Taylor & Francis, Boca Raton, 4th edn, 2013.

11 Plant–Insect Interactions and the Human Role

11.1 Introduction

The previous chapter addressed interactions between insects and plants that crucially affect humans. Insects that pollinate vital food crops must be encouraged and protected in order to feed a world with an ever-increasing human population. On the other hand, some insects regularly destroy vast acreages of croplands and infest large stands of timber. As they severely reduce the yields of required plant products, they must be controlled.

This chapter describes some examples of herbivorous insects that at times infest and greatly damage plants in agriculture and horticulture. It then addresses different approaches to control undesirable insects. The challenge is to design effective methods that manage the insect pests while minimizing detrimental effects on beneficial insects and other organisms, as well as on the natural environment. Some examples of synthetic insecticides are shown, with their development and their historic uses. The chapter concludes with thoughts on the management of insect pests on essential crop plants while attempting to safeguard the environment and biodiversity.

11.2 Insects That Damage Crop Plants

In agriculture and horticulture, an infestation of insects on a crop is considered a pest if it causes enough damage to cause economically

The Chemistry of Plants and Insects: Plants, Bugs, and Molecules
By Margareta Séquin
© Margareta Séquin 2017
Published by the Royal Society of Chemistry, www.rsc.org

important losses.[1,2] Therefore, the concept of an insect 'pest' is entirely conceived from the human point-of-view. Amazingly, less than one percent of insect species worldwide are pests, and only a few hundred of these species consistently cause serious problems. Following are a few examples of insects that have been highly damaging to crops. The European corn borer (*Ostrinia nubilalis*) on corn or maize (*Zea mays*) can be most destructive on maize fields in North America. The cotton boll weevil (*Anthonomus grandis*), an insect originally from Mexico, attacks cotton (*Gossypium* sp.) plantations in North and South America. The Colorado potato beetle (*Leptinotarsa decemlineata*) damages potato plants (*Solanum tuberosum*) all over the world. Masses of different species of aphids infest many different crops, especially legume plants (Fabaceae).[3] Gypsy moths and various bark beetles cause vast damage in forests (see Chapter 6.1). Fruit flies (family Tephritidae), especially the Mediterranean fruit fly (*Ceratitis capitata*), inflict severe losses on more than two hundred species of fruits and vegetables. While crop losses may be a nuisance for the home gardener, they can be a matter of survival for farmers of large acreages.

A formerly harmless insect can become a pest for different reasons. Insects imported from a different part of the world may encounter no natural enemies, that exert a controlling influence on them, in a new environment. The lack of predators can enable nonnative insects to rapidly multiply and proliferate. (The common names of damaging insects often allude to their nonnative origins.) Native insects may also adapt to feeding on introduced nonnative plant species, as was the case of the formerly oligophagous Colorado potato beetles that now specifically feed on potato plants (Chapter 1). International trade has helped to accidentally introduce insects that damage local crop plants.

The development of insect pests on crops is supported by the prevalence of raising monocultures, in which a single species of crop plants or forest trees is grown in dense stands of identical plants. Monocultures provide ease of harvesting and, under ideal conditions, deliver large yields of crops. But when invaded by specialist or generalist insects, monocultures provide a bonanza of feed for the insects, making the crops vulnerable to vast damage. The lack of diversity of plants in a monoculture reduces the development of a diversity of insects, including natural predators of the pest insects. Therefore, invasive herbivorous insects are less likely to be kept in check.

11.3   Approaches to Managing Insect Pests

All through history humans have had to contend with insect infestations of plant crops. Swarms of locusts destroyed entire harvests
during biblical times and led to periods of human starvation. Insects
can rapidly evolve through short sequences of generations and
therefore can quickly adapt to methods applied to control them.
Managing the invasions of damaging insects on crop plants, to a level
that makes losses tolerable, will be an ongoing task for humans.[4]

Ideally, the control, or rather the management, of damaging insects
involves methods that successfully limit the unwanted pests, but do
not harm beneficial insects or other organisms (including humans),
or damage soils and waterways. The control methods and media
should be economically viable and relatively simple to apply. As desirable as such solutions might be, they are difficult to find. As insects
develop resistance to methods of management, new approaches must
be developed and constantly adjusted and redesigned. Climate
change further challenges methods that are developed. Various
approaches that have been applied or that are in use are described in
the following parts of this chapter.

Historically, people have used the practice of alternating crops,
including rotations with plants that have nitrogen-fixing bacteria to
retain soil fertility. The diversity of plantings supported the diversity
of insects and kept damaging insects in check. While these
approaches to insect management may be adequate for smaller scale
plantings, the pressure to successfully and consistently produce large
volumes of crop products makes the approach of crop rotations
economically challenging and often unacceptable.

In biological control, predators that are natural enemies of herbivorous insects are imported to feed on the insect pests and thus keep
economic losses of a crop to a tolerable level. Ladybird beetles as
predators on aphids are well-known to home gardeners.

The use of pheromones in mass-trappings has been successful,
especially in forestry and orchards, as traps with highly specified
pheromones can be aimed at specific pests. They are less useful in
very large-scale plantings. Bad weather conditions make pheromone
traps useless. The timing of setting out the traps is crucial and
therefore requires detailed knowledge of the life cycles of the pest
insects for successful management. Pheromone traps have been used
with good success to lure and combat bark beetles, especially of the
genus *Ips*.[5] (Compare Chapter 5.2.)

Present research efforts strongly focus on the development of genetically engineered plants, *i.e.* transgenic plants, that are resistant to an insect pest.[6] In these techniques, genes encoding proteins from other plants or from bacteria, commonly from *Bacillus thuringiensis* (*Bt*), are inserted into the genome of a crop plant. The selected genes express toxins that are specifically aimed *e.g.* at certain orders of insect pests, like larvae of butterflies and moths that damage forests or vegetable and field crops. The genetically engineered seeds that produce these crops have continuous protection from invading insects. Therefore, the application of insecticides can be avoided. Concerns focus on the gradual development of resistance of the pest insects towards the defensive substances expressed in the crop plants. Currently, legislation strictly regulates the use of genetically engineered plants.

At this time, the use and application of chemical insecticides is the most commonly used method to manage insect pests. The following descriptions address classes of commonly used synthetic insecticides, including some of their history.

11.4 Synthetic Insecticides

Some of the insecticides used to control insect pests are natural plant-derived products, like the pyrethrins shown in the previous chapter. Others are synthetic products, with chemical structures that are derived from natural products. Yet other insecticides have no natural analogs. Most synthetic insecticides are broad spectrum in action and target the insects' nervous systems.

The potent insecticidal efficiency of DDT (dichlorodiphenyltrichloroethane) was discovered by Paul Hermann Müller in 1939. (He received a Nobel Prize for his work in 1948.) The insecticide was used in the second half of World War II to combat malaria and typhus. After the war, DDT was extensively applied worldwide on plants as an agricultural insecticide. But what looked like the "ideal" insecticide became known for its widespread damaging environmental effects, induced by uncontrolled use of DDT, as described in Rachel Carson's book "Silent Spring".[7] DDT **11.1** (Figure 11.1) is a chlorinated hydrocarbon, *i.e.* a hydrocarbon in which several of the hydrogen atoms are replaced by chlorine atoms. DDT is lipophilic and accumulates in the environment and biological systems. It is a persistent insecticide in food chains. The agricultural use of DDT is now banned worldwide.

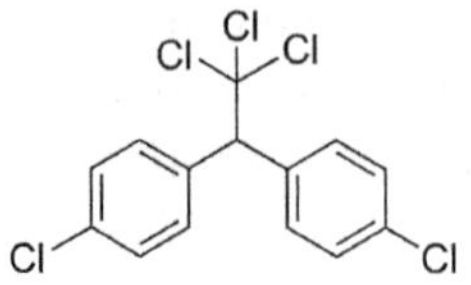

Figure 11.1   Examples of synthetic insecticides. DDT (dichlorodiphenyltri-chloroethane) **11.1** is a chlorinated hydrocarbon. Malathion **11.2** is an organophosphate. Imidacloprid **11.3** is an example of a neonicotinoid.

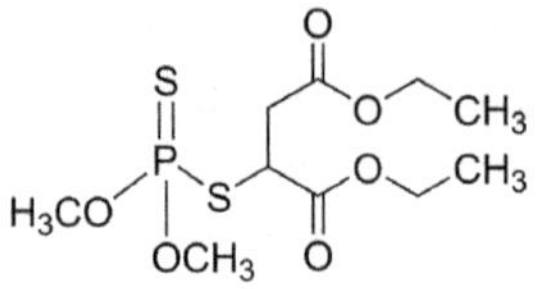

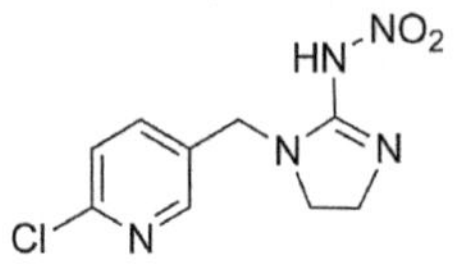

Organophosphates are a class of synthetic pesticides that are widely used in agriculture and in residential landscaping. An example of a commonly used organophosphate is malathion **11.2**. While it is considered to have low toxicity towards mammals (including humans), it is toxic to bees and other beneficial insects, and also to some fish and other aquatic life.[8]

The use of neonicotinoids, like imidacloprid **11.3** (Figure 11.1), is widespread. The chemical structures of neonicotinoids have some resemblance to the alkaloid nicotine **10.29** (Figure 10.14), itself a potent insecticide. In recent years there has been increasing evidence that honey bee colony collapse disorder may be connected with the use of neonicotinoids.[9]

Insecticides tend to be highly successful in suppressing insect pests initially and contribute to the production of large harvests of crops. But their side-effects include the killing of non-target organisms, like pollinators, natural enemies of pests, and beneficial soil organisms, as well as potential toxicity towards higher organisms. Insecticides can promote the selection of pest insects that are genetically resistant to them. The timing of insecticide application is essential; it has to occur at a vulnerable stage of an insect pest. Dosage must be kept at minimum amounts required. While much research has been done and best strategies for application are available, education on the proper use of pesticides is essential.[10]

Integrated pest management (IPM) is a sustainable approach to managing pests that combines different techniques, including biological, physical, chemical, and other control methods and that attempts to minimize health and environmental risks. Successful integrated pest management requires thorough interdisciplinary knowledge of the biology and ecology of the systems involved, of the insects, of other animals, and of the respective habitats. It requires the necessary training of people involved in the application of the recommended methods for best success, which is time consuming, costly, and often unavailable.[11]

11.5 Conclusions

The rapid evolution of insects allows their quick adaptation to chemical compounds in plants, natural or artificial. Humans, in need of producing adequate quantities of food, have developed powerful methods to control herbivorous insects on food crops. While applying the controls to the unwanted insects, other organisms, as well as the environment in general, have been frequently damaged.

The search continues for successful and improved methods of managing insect pests, in order to produce adequate food crops and to protect stands of timber. At the same time these efforts must be integrated with environmentally benign methods, in order to protect ecosystems and to conserve biodiversity.

References

1. P. J. Gullan and P. S. Cranston, *The Insects: An Outline of Entomology*, Wiley Blackwell, Chichester, West Sussex, 4th edn, 2010, ch. 16, pp. 408–441.
2. L. M. Schoonhoven, J. J. A. van Loop and M. Dicke, *Insect-Plant Biology*, Oxford University Press, Oxford, 2005.
3. R. L. Blackman and V. F. Eastop, *Aphids on the World's Crops: an Identification and Information Guide*, Wiley, New York, 2000.
4. K. C. Kim, Insect Pests and Evolution, in *Evolution of Insect Pests/ Patterns of Variation*, ed. K. C. Kim and B. A. McPherson, John Wiley & Sons, Inc., 1993, ch. 1.
5. F. Schlyter, Q. Zhang, G Liu and L Ji, A successful Case of Pheromone Mass Trapping of the Bark Beetle *Ips duplicatus* in a Forest Island, *Integr. Pest Manage. Rev.*, 2001, **6**(3), 185.
6. S. O. Duke, Comparing Conventional and Biotechnology-Based Pest Management, *J. Agric. Food Chem.*, 2011, **59**, 5793.
7. R. Carson, *Silent Spring*, Houghton Mifflin, New York, NY, 1962.
8. S. E. Manahan, *Fundamentals of Environmental and Toxicological Chemistry*, CRC Press, Taylor & Francis Group, Boca Raton, FL, 4th edn, 2013.

9. R. Nauen and I. Denholm, Resistance of insect pests to neonicotinoid insecticides: Current status and future prospects, *Arch. Insect Biochem. Physiol.*, 2005, **58**(4), 200.

10. *Current Controversies Series: Pesticides*, ed. D. A. Miller, Greenhaven Press, MI, 2014.

11. L. E. Ehler, Integrated pest management (IPM): definition, historical development and implementation, and the other IPM, *Pest Manage. Sci.*, 2006, **62**, 787.

Epilogue

It has been the aim of this book to introduce some of the fascinating chemistry involved in the communications between plants and insects. Plants have evolved a huge number of diverse organic compounds that support their survival. Insects, often in response to plant substances, have developed a wealth of different compounds and mechanisms that enable them to use plants as food and sometimes as defense. The illustrated examples in this book have introduced general characteristics of organic compounds in this context. They demonstrate the central role of chemistry in natural processes.

Many interactions between insects and plants are of great interest to people, whether they are desirable or doing great damage. We hear about insect infestations of forests and crop plants or about threats to pollinators. Projects at the forefront of research include diverse studies of communications between plants and insects. Investigations may be aimed at gaining insight into the communication mechanisms or at developing ever-improving methods to manage invasive insects. They may also focus on determining threats to the environment by the control methods. An understanding of the chemistry background can help us to critically evaluate the different approaches used to manage these interactions. It is hoped that this book encourages the reader to observe insects and plants and to reflect on their communications in the natural world.

The Chemistry of Plants and Insects: Plants, Bugs, and Molecules
By Margareta Séquin
© Margareta Séquin 2017
Published by the Royal Society of Chemistry, www.rsc.org

Glossary

How to Read Structures of Organic Compounds: a Brief Guide

The structures of organic compounds are most commonly shown as line structures or bond-line structures in this text. A line represents an electron pair of a covalent bond. An angle signifies a carbon atom. A line ending represents a –CH_3 group. For hydrogen atoms that are not shown, add enough hydrogens in your mind to provide each carbon atom with its required four bonds. Atoms other than carbon and hydrogen (O, N, S *etc.*) are always written out. Refer to chemistry texts for more details if needed (see Chapter 1).

How to Understand Systematic Plant and Insect Names

The binomial system of plant names and insect names is based on Carl Linnaeus' work. It assigns a unique name to each species of plant or insect which universally defines a species, unlike common names which can have a lot of variations. Systematic names consist of genus (pl. genera) and species and are usually italicized.

Example: *Asclepias speciosa.* Genus: *Asclepias*, species: *speciosa*. Common name: Showy milkweed. Another species: *Asclepias californica*.

A similar system is used to name insects.

Example: *Danaus plexippus.* Genus: *Danaus*, species: *plexippus*.

The Chemistry of Plants and Insects: Plants, Bugs, and Molecules
By Margareta Séquin
© Margareta Séquin 2017
Published by the Royal Society of Chemistry, www.rsc.org

Common names: Monarch butterfly, or monarch.

Abbreviation of species: sp., which is also used when a species is not defined.

In the hierarchy of systematic naming genera are grouped into families, and families are grouped into orders.

Glossary of Terms

Aglycon: The non-sugar portion of a glycoside.

Aliphatic compounds: Organic compounds that do not contain an aromatic system.

Alkaloids: A large family of nitrogen-containing, organic natural products, mostly from plants, with basic (alkaline) properties.

Angiosperms: Flowering plants.

Anthocyanins: A large group of water-soluble phenolic plant pigments. A subgroup of flavonoids.

Aposematic colors: Warning colors. Distinct, contrasting coloring or color patterns, especially of insects, that alert predators of the insect's unpalatability.

Aqueous: Dissolved in water.

Autotrophs: Organisms that are able to synthesize their nutritive organic substances from simple inorganic substances in their environment.

Auxins: Plant hormones that actively promote the growth of plant cells.

Betalains: A family of nitrogen-containing plant pigments that are only found in specific plant families.

Biosynthesis: The biological synthesis of natural products.

Carbon skeleton: The carbon–carbon backbone of an organic molecule.

Cardenolides or **cardiac glycosides:** A group of steroid glycosides from plants that affect the activity of the heart muscle.

Carotenoids: A group of isoprenoid pigments with 40 carbon atoms. Tetraterpenoids.

Cecidology: The study of plant galls.

Chiral: Asymmetric or 'handed'.

***Cis/trans* isomers:** Stereoisomers that differ in the positioning of atoms attached to a carbon–carbon double bond (or a ring).

Coevolution: Close interactions between two or more organisms, resulting in strong selective forces on each other and leading to evolutionary adaptations.

Conformations: Spatial arrangements of the atoms in a molecule obtained by free rotation around single bonds.

Conjugated double bonds: Alternating single bond-double bond sequences.

Cyanogenic glycosides: A group of plant glycosides that release HCN when reacting with enzymes from neighboring plant cells.

Disaccharides: Carbohydrates consisting of two sugar units. An example is sucrose.

Diterpenes: Terpenes consisting of four isoprene units and thus having twenty carbon atoms.

Domatia: Small, chamber-like structures on plants that serve as hiding places for ants.

Enantiomers: Compounds that are mirror images of each other.

Essential amino acid: An amino acid that cannot be produced by an organism itself and thus has to be ingested through nutrition. Different organisms may have different sets of essential amino acids.

Essential oils: Volatile, fat-soluble plant oils (from the word "essence").

Extrafloral nectaries: Plant structures that provide nectar but that are not associated with flowers.

Exudate: A fluid or sap that is formed as a response to injury.

Flavonoids: A large family of plant pigments with a common phenolic three-ring structure.

Glucosinolates: A family of sulfur-containing, defensive plant glycosides.

Glycoside: A molecule that consists of a non-carbohydrate part (the aglycon) and a sugar part.

Gum: Water-soluble plant exudate that consists of large carbohydrate structures, usually formed as a response to injury.

Herbivores: Organisms that feed on plants.

Heterotrophs: Organisms that cannot manufacture organic compounds and therefore have to obtain basic organic nutrients from other organisms (namely from autotrophs).

Honeydew: A sugary solution excreted by insects like aphids.

Hydrolysis: Cleaving a molecule by reaction with water.

Hydrolyzable tannins: Water-soluble, phenolic plant compounds derived from gallic acid.

Hydrophilic: Having an affinity for water. (Opposite of hydrophobic.)

Hydrophobic: Water-repelling.

Instar: Phase of development of an insect. The growth stage between two successive molts.

Iridoids: Chemical insect defenses related to monoterpenes.

Isomers: Molecules with the same molecular formula but with different structures.

Isoprene unit: Characteristic C_5 building block of terpenes.

Isoprenoids: Another name for terpenes.

Latex: A milky, aqueous plant sap that contains rubber particles.

Lipid: Any non-polar substance in biological systems that is fat-soluble, but insoluble in water.

Lipophilic: Fat-soluble, water-insoluble.

Metabolism: The sum of all chemical reactions occurring in living organisms.

Molecular formula: A formula that shows the total number of atoms of each element in a molecule. Example: $C_6H_{12}O_6$.

Monomers: The subunits that serve as building blocks in polymers.

Monophagous insects: Insects that feed on one type of plant.

Monosaccharides: The simplest carbohydrates or sugars, consisting of one sugar unit. Example: glucose.

Monoterpenes: Terpenes that consist of two isoprene units and thus have ten carbon atoms.

Mucilage: Gel-like, viscous, hydrophilic mixture of polysaccharides and other polymers, found in many plants.

Multitrophic interactions: Interactions that involve several trophic levels in a food web, *e.g.* between plants and insects.

Mutualistic interactions: Biological interactions that are beneficial to all organisms involved.

Myrmecophylous: Ant-loving, ant-attracting.

Myrmecophytes: Ant plants. Plants that have beneficial relationships with ants.

Nectar guides: Pigment designs in flowers that lead pollinating insects to nectaries.

Nectary: A gland that secretes nectar in plants.

Neurotransmitters: Compounds that act as chemical messengers within the nervous system.

Oligophagous insects: Insects that feed on a limited number of different plants.

Peptide: A molecule that consists of two to about a hundred amino acid monomers or more. Not as large as proteins.

Phenolics: Organic molecules that contain phenolic rings, *i.e.* aromatic rings with hydroxy groups (OH) attached to them.

Pheromone: Organic compound emitted by an animal to communicate with another animal of the same species and that affects its behavior, *e.g.* sexually, as an alarm, or for trail following.

Phytophagous: Feeding on plants (from Greek: phyto for plant; phagous for eating).

Pigment: A compound that absorbs a section of the electromagnetic spectrum of sunlight and reflects or transmits the remaining wavelengths.

Polymers: Giant molecules that consist of a large number of repeating subunits (monomers).

Polyphagous insects: Insects that feed on many different plants.

Polysaccharides: Polymeric carbohydrates. Examples: starch, cellulose.

Proteins: Polymers that consist of amino acid monomers linked by peptide bonds.

Pterins: Pigments with heterocyclic structures found in butterfly wings.

Resin: A water-insoluble, elastic plant exudate.

Sesquiterpenes: Terpenes consisting of three isoprene units, thus having fifteen carbon atoms.

Stereoisomers: Molecules with the same connections of atoms but with different orientation of groups in space. Examples: enantiomers, *cis/trans* isomers.

Steroids: A large family of organic natural products with a characteristic four-ring system.

Stigma: Part of a flower that receives pollen from pollinators.

Tannins: Acidic, phenolic plant pigments with astringent properties.

Terpenes: Secondary metabolites composed of isoprene units.

Total synthesis: Synthesis of a complex compound from simple, easily-available organic compounds, accomplished in the laboratory.

Vacuole: A space or cavity bounded by membranes within the cytoplasm of a cell, filled with a watery fluid.

Vesicant: A chemical compound that causes severe blistering, chemical burns, and other tissue injury, potentially leading to tissue necrosis (tissue death).

Volatile: Evaporating easily.

Xanthophylls: Yellow plant pigments related to carotenoids.

Subject Index

Page references to *figures and tables* are shown in *italics*.
Plants and insects have been indexed under both their scientific and common names unless the names are very similar, in which case only the scientific name is used in the index.